# Soil and Water Chemistry

## *Theory Manual*

## THE AUTHORS

**Dr. M. Divya** is an Assistant Professor and Head, Department of Aquatic Environment Management at the College of Fisheries Science, St. Devasahayam Institute of Fisheries Science and Technology, located in Midalam, Kanniyakumari District, Tamil Nadu. With four years of research experience under UGC and DBT funded projects and two years of teaching experience, she has contributed significantly to her field. She has published over 15 research papers in reputable national and international journals. Additionally, she conducted a capacity-building program funded by NFDB, Hyderabad, and has actively participated and presented in numerous national and international conferences, symposiums, workshops, and seminars. Her expertise lies in the fields of planktonology and bioremediation technology.

**Prof. S. Felix** is a distinguished figure in the field of fisheries education and technology consulting, with a career spanning over four decades. Currently serving as the Dean of the College of Fisheries Science, St. Devasahayam Institute of Fisheries Science and Technology in Midalam, Kanniyakumari district, Tamil Nadu and also President of AQUAFIN (Aquaculture Technology & Research Foundation India), he has held several prestigious positions, including Vice Chancellor of Tamil Nadu Fisheries University (TNJFU) and President of the World Aquaculture Society (WAS)-APC. Prof. Felix is recognized for his exemplary leadership and contributions to the fisheries and aquaculture sector, demonstrated through his involvement in various committees and organizations. He is a prolific author, having published numerous research papers, books, and manuals, and has played a pivotal role in introducing new fisheries science courses and establishing fisheries colleges and research centers.

As per ICAR's 5th Deans committee Recommendation

# Soil and Water Chemistry

## *Theory Manual*

**M. Divya**
**S. Felix**

College of Fisheries Science
St. Devasahayam Institute of
Fisheries Science & Technology
Midalam, Kanniyakumari Dist.629193 (Tamil Nadu)

2025

Daya Publishing House®
*A Division of*
Astral International Pvt. Ltd.
New Delhi – 110 002

ISBN: 978-93-5919-310-6



*Published by* : **Daya Publishing House®**
*A Division of*
**Astral International Pvt. Ltd.**
– ISO 9001:2015 Certified Company –
4736/23, Ansari Road, Darya Ganj
New Delhi-110 002
Ph. 011-43549197, 23278134
E-mail: info@astralint.com
Website: www.astralint.com

*Printed at* : **Replika Press Pvt. Ltd**

# Preface

Welcome to this manual on soil and water chemistry. As the importance of environmental sustainability continues to grow, understanding the intricate dynamics of soil and water has become increasingly crucial. This manual aims to provide readers with a comprehensive overview of these essential environmental elements, offering insights into their chemical processes and interactions.

In developing this manual, our goal was to create a resource that is accessible yet informative, catering to a wide range of readers, from students to professionals in the field. We have strived to present complex concepts in a clear and concise manner, supported by practical examples and case studies.

Each chapter of this manual is dedicated to exploring different aspects of soil and water chemistry, from the fundamental principles to practical applications. We delve into topics such as soil composition, water quality criteria, soil and water quality criteria for aquaculture practices.

We hope that this manual will serve as a valuable resource for anyone seeking to deepen their understanding of soil and water chemistry, whether for academic study, professional development, or environmental stewardship. We also encourage readers to engage with the material, ask questions, and explore further avenues of research.

Finally, we would like to express our gratitude to all the contributors who generously shared their knowledge and expertise to make this manual possible. Without their dedication and collaboration, this manual would not have been realized.

We hope you find this manual informative and inspiring, and we invite you to join us in our journey to better understand and protect our precious soil and water resources.

**M. Divya**

**S. Felix**

# Contents

Chapter-1

# Analytical Chemistry: Principles, Applications and Types

## Introduction

Analytical chemistry is the branch of chemistry that deals with the analysis of different substances. Analytical chemistry involves the separation, identification, and quantification of matter. It involves the use of classical methods along with modern methods involving the use of scientific instruments.

Analytical chemistry involves the following methods:

- The process of separation isolates the required chemical species which is to be analysed from a mixture.
- The identification of the analyte substance is achieved via the method of qualitative analysis.
- The concentration of the analyte in a given mixture can be determined with the method of quantitative analysis.

## Branches of Analytical Chemistry

Two sub-branches come under analytical chemistry namely quantitative analysis and qualitative analysis which can be explained as follows.

### Quantitative Analysis

Quantitative analysis is a method of determining the absolute or relative quantity regarding the concentration of one or more substances present in a sample or compound. Some quantitative analysis techniques include Gravimetric analysis and Volumetric analysis.

### Qualitative analysis

Quality means the standard or the feature of one substance. Hence, Qualitative analysis method deals with the determination of the quality of a particular compound, irrespective of its quantity or concentration. In simpler words, the qualitative analysis does not measure the amount of the substance but measures the quality of that material. One of the best examples of this type of method is the observation of a chemical reaction, whether there will be a change in colour or not. The qualitative analysis method can be measured in different ways such as chemical tests, flame tests, etc. Such tests are widely used in salt analysis (identification of the cation & anion of inorganic salts).

## Methods Used in Analytical Chemistry

The methods used to determine the identity and the quantity of the analytes in the field of analytical chemistry can be broadly divided into classical and instrumental methods.

### Classical Methods

These methods can further be divided into

- Separation methods
- Identification methods

### 1. Separation methods

- **Precipitation:** This is a gravimetric method in which the reactants and products of a chemical reaction are used to analyse a substance. The metallic ions of many elements may enter into a reaction with the negative ions to form a new insoluble substance called the precipitate, which settles to the bottom of the solution. This is filtered out and washed.
- **Extraction:** This is a procedure in which a substance is removed from a matrix, both being in two immiscible phases. The most common technique for extraction of a compound from an aqueous solution is liquid-liquid extraction (LLE). Solid phase extraction (SPE) is a widely used method of sample preparation used to separate and enrich purified components from aqueous solution.
- **Distillation:** This is the process of evaporation and condensation by which a component is separated out of a mixture of liquids. It is based upon the difference in boiling points or volatility between the various substances present in the mixture.

### 2. Identification methods

These methods are based on colour, odour, melting and boiling points, radioactive or reactive properties of the analyte.

### A. Classical Quantitative Methods

- **Titration method:** It includes acid-base or complexometric titration. Acid-base titration helps in quantifying an unknown acid or base in solution by finding its concentration based on the exact amount of corresponding base or acid required to neutralize it. Complexometric titration is a volumetric analysis where end point of titration is the formation of a characteristic-coloured complex, which is usually a metal complex.
- **Gravimetric method:** It uses precipitation or volatilization to determine the mass of the analyte from its ionic mass. Precipitation, volatilization or electro analytical methods are mostly used.

### B. Instrumental methods:

Instrumental methods are mostly used in advance level of research and the same instrument may be used to separate, identify and quantify the analyte.

- **Chromatography:** In this process, the mixture to be separated on a stationary phase (solid or liquid) and a pure solvent such as water or any gas is allowed to move slowly over the stationary phase, carrying the components separately as per their solubility in the pure solvent.
    - Paper or thin layer chromatography
    - Gas chromatography
    - Ion-exchange and size-exclusion chromatography
    - Reversed-phased chromatography
    - Two-dimensional chromatography
    - Hydrophobic interaction chromatography
- **Potentiometry:** It is a method of analysis in which the concentration of an ion or substance is determined by measuring the potential developed when a sensitive electrode is immersed in the solution of the species to be determined. Commonly available pH meters are designed based on this principle.
- **Conductometry:** Conductometry is the measurement of the electrical conductivity of a solution. The main principle involved in this method is that the movement of the ions creates the electrical conductivity. The movement of the ions is mainly depended on the concentration of the ions. Electrical conductivity and Total Dissolved Solids (TDS) of water samples generally determined by instruments designed based on this principle.
- **Turbidimetry/ Nephelometry:** The principle of nephelometry and turbidimetry is based on the on the well-known law of Tyndall Effect. When light is passed through the suspension, part of incident radiant energy is dissipated by absorption, reflection, and refraction while reminder is transmitted. Measurement of the intensity of transmitted light is a function of the concentration of dispersed phase and this becomes the basis of turbidimetric analysis.

- **Electrophoresis:** In this method, an electric field is used to separate particles in a fluid because of development of a charged interface between them.

### *C. Instrumental qualitative and quantitative methods:*

These methods may utilize light, electromagnetic, heat and combinations of these to identify the analyte.

- **Spectrometry:** This method can isolate particles based on various properties like mass, momentum or energy. It includes techniques such as ion-mobility spectrometry based on mobility of ionised particles in a carrier gas, or mass spectrometry which measures the ratio of mass to charge of ions.
- **Spectrophotometry:** This method is based on Beer-Lambert law. Here a material or component is identified by its reflection or transmission of light, which depends on the wavelength. Visible, ultraviolet or near-infrared light may be used in this method. This method is widely used to study gases, solids or solutions.

**Beer-Lambert Law**: This law states that the absorbance of a light absorbing material is proportional to its concentration in solution. The general Beer-Lambert Law is usually written as

$$A = elc$$

*Where,*

A - Measured absorbance

e - Extinction coefficient of the substance- unique for each substance

$l$ - Sample path length measured in centimetres (i.e. the width of the cuvette almost always 1 cm)

c - Molar concentration of the solution (expressed in terms of molarity)

- **Radiometry:** Radiometry is the science of measuring light in any portion of the electromagnetic spectrum. In practice, the term is usually limited to the measurement of infrared, visible, and ultraviolet light using optical instruments. Irradiance is the intensity of light and is measured in watts per square meter.

### *D. Volumetry & Gravimetry*

- **Volumetry:** It is a widely used quantitative analytical method. It involves measurement of volume of a solution of known concentration which is used to find out the concentration of the analyte.

  Some of the widely used volumetric analysis are:

  - **Acidimetry:** The term acidimetry refers to that part of volumetric analysis whereby an acid solution at known concentration, along with a specific indicator, is used to titrate a base solution and thus work out its concentration.

- ☆ **Alkalimetry:** The term alkalimetry refers to that part of volumetric chemical analysis which enables us to work out the concentration of an acid solution using an alkaline solution at a known concentration and a suitable indicator.

- ❑ **Gravimetry:** Gravimetric analysis is a technique through which the amount of an analyte (the ion being analysed) can be determined through the measurement of mass. Gravimetric analyses depend on comparing the masses of two compounds containing the analyte.

## Applications of Analytical Chemistry

Some important applications of this branch of chemistry are listed below.

- ❑ The shelf lives of many medicines are determined with the help of analytical chemistry.
- ❑ It is used to check for the presence of adulterants in drugs.
- ❑ Soil can be tested to check for appropriate concentrations of minerals and nutrients that are necessary for plant growth.
- ❑ It is employed in the process of chromatography where the blood samples of a person are classified.
- ❑ The concentration of the pesticide residues and the contaminants in a given food sample can also be determined via analytical chemistry.
- ❑ It also has many important applications in medicine, with its use in the testing of cholesterol and glucose levels in a blood sample.
- ❑ Analytical chemistry is an integral part of forensic science, clinical analysis, and even environmental analysis.

## Standard Solutions

Standard solutions are solutions of accurately known concentrations, prepared using standard substances. There are two types of standard solutions known as primary solution and secondary solution. A primary standard solution is a solution with a high purity and less reactivity. A secondary standard is not that pure and is chemically reactive than primary standards.

**Difference between primary and secondary standard solution**

| Primary Standard solution | Secondary standard solution |
|---|---|
| Primary standard solutions are solutions made out of primary standard substances | Secondary standard solutions are solutions made specially for a certain analysis |
| Extremely pure (About 99.9%) | Not very pure |
| Less or not reactive | Comparatively more reactive |
| Not hygroscopic | Somewhat hygroscopic |
| Used to standardize secondary standards and other reagents | Used for specific analytical experiments |

# Titration & Indicators

## Titration:

- Titration is a laboratory technique used to determine the concentration of a solution by reacting it with a solution of known concentration.
- It involves slowly adding the known solution (titrant) to the solution of unknown concentration (analyte) until a reaction is complete.
- The point at which the reaction is complete is called the equivalence point.
- Various types of titrations exist, including acid-base, redox, complexometric, and precipitation titrations.

**Different types of titrations**

| Titration Type | Description | Example |
|---|---|---|
| Acid-Base Titration | Involves the neutralization reaction between an acid and a base. The endpoint is indicated by a pH indicator. Produces water and a salt. | Titration of hydrochloric acid (HCl) with sodium hydroxide (NaOH) using phenolphthalein as an indicator. The endpoint is reached when the solution changes from colourless to pink, indicating complete neutralization. |
| Redox Titration | Involves oxidation-reduction reactions. Classified into dichromate, permanganate, iodometric, and iodimetric titrations, depending on the oxidizing agent used. Endpoint may be indicated by a colour change or other methods. | Titration of iron (II) sulfate ($FeSO_4$) with potassium permanganate ($KMnO_4$) in acidic solution. The endpoint is reached when the purple colour of permanganate solution persists, indicating the oxidation of $Fe^{2+}$ ions to $Fe^{3+}$ ions. |
| Complexometric Titration | Involves the formation of stable complexes between metal ions and complexing agents. EDTA (Ethylene diamine tetra acetic acid) is commonly used. Endpoint often determined by metal ion indicators or by the appearance of a precipitate. | Titration of calcium ions ($Ca^{2+}$) with EDTA using Eriochrome Black T as an indicator. The endpoint is reached when the solution changes colour from wine-red to blue, indicating the formation of a stable complex between calcium ions and EDTA. |
| Precipitation Titration | Involves the formation of a precipitate as a result of a chemical reaction between the analyte and titrant. Endpoint is determined by the appearance of a precipitate. Commonly used to determine halides, sulphates, and other ions. | Titration of chloride ions ($Cl^-$) with silver nitrate ($AgNO_3$) using potassium chromate ($K_2CrO_4$) as an indicator. The endpoint is reached when the appearance of a red-brown precipitate of silver chromate ($Ag_2CrO_4$) indicates the complete reaction between chloride ions and silver ions. |

## Indicators

- Indicators are substances that change colour in response to a change in the conditions of the solution, such as pH.
- They are used to signal when the reaction between the titrant and analyte is complete.
- Common indicators include phenolphthalein, methyl orange, bromothymol blue, and litmus.
- The choice of indicator depends on the pH range of the titration and the nature of the substances being titrated.
- Indicators must be selected carefully to ensure their colour change corresponds accurately to the equivalence point of the titration.

## Types of Indicators

- **Acid base indicators**
  - ☆ Used in neutralization reactions. Change their colour depending on pH of solution. e.g. Phenolphthalein, Methyl red, litmus etc.
- **Redox indicators**
  - ☆ Used in oxidation – reduction reactions. Colour changes with the change of electromotive force e.g., $KMnO_4$, N-phenyl anthranilic acid.
- **Adsorption indicators**
  - ☆ Change of colour due adsorption of a substance on its surface.
- **Radioactive indicators**
  - ☆ Used in the study of variety of reactions as tracer. Their presence is detected by suitable devices. e.g., $D^2$, $O^8$,$I^{127}$etc.

## Selection of Indicators

- The choice of indicator depends on the pH range over which the titration is performed.
- Indicators should have a sharp and distinct colour change at or near the equivalence point of the titration.
- The indicator's colour change should be easily observable and easily distinguishable from the colour of the solution before titration.

## Accuracy and Precision

- ☆ The accuracy and precision of a titration depend on the correct selection and use of indicators.
- ☆ If the indicator's colour change does not correspond accurately to the equivalence point, the titration results may be inaccurate.
- ☆ Proper technique, careful observation, and calibration are essential to ensure accurate and precise titration results.

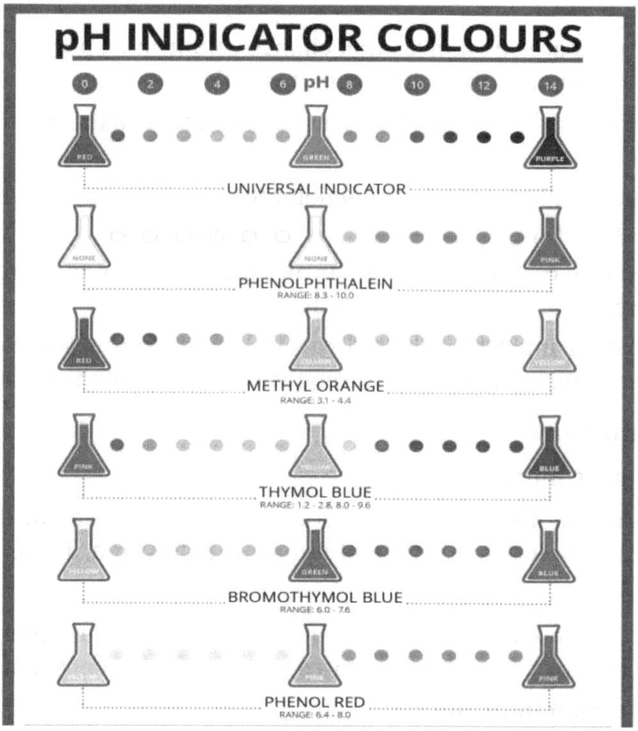


**Commonly used acid-base indicators**

## Dilute Solutions

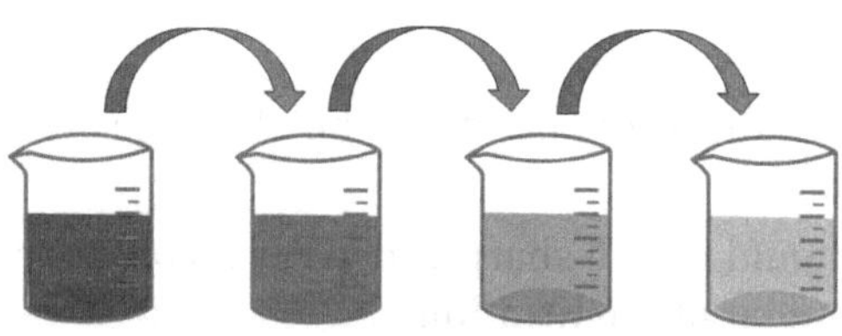

Dilution is the process of adding additional solvent to a solution to reduce its concentration. This process keeps the amount of solute constant, but increases the total amount of solution, thereby decreasing its final concentration. Dilution can be done by mixing a solution of higher concentration with an identical solution of lesser concentration. Diluting solutions is a routine process in the laboratory, as stock solutions are often purchased and stored in very concentrated forms. For the solutions to be usable in the lab (for a titration, for instance), they must be accurately diluted to a known, lesser concentration. The volume of solvent needed to prepare the desired concentration of a new, diluted solution can be calculated mathematically. The relationship is as follows:

$M_1V_1 = M_2V_2$

Where,

$M_1$ - concentration of the original solution

$V_1$ - volume of the original solution

$M_2$ - concentration of the diluted solution

$V_2$ - final volume of the diluted solution

When calculating dilution factors, it is important that the units for both volume and concentration are the same for both sides of the equation.

**Example:**

A 10L solution of 5M HCl is diluted to 1M HCl solution. What is the volume of the 1M HCl solution formed?

$M_1 = 5M, M_2 = 1M, V_1 = 10L, V_2 = ?$

$M_1V_1 = M_2V_2$

$5 \times 10 = 1 \times V_2$

$V_2 = 50$ L

## Serial Dilutions

Serial dilutions involve diluting a stock or standard solution multiple times in a row. Typically, the dilution factor remains constant for each dilution, resulting in an exponential decrease in concentration. As for example, a ten-fold serial dilution could result in the following concentrations: 1 M, 0.1 M, 0.01 M, 0.001 M, and so on. As seen in this example, the concentration is reduced by a factor of ten in each step.

Serial dilutions are used to accurately create extremely diluted solutions, as well as solutions for experiments that require a concentration curve with an exponential or logarithmic scale. Serial dilutions are widely used in experimental sciences, including biochemistry, pharmacology, microbiology, and physics.

## Units of Concentration

Most commonly, a solution's concentration is expressed in terms of normality, molarity, molality, formality etc.

### A. Normal solution

A normal solution contains one-gram equivalent mass of the substance in one litre of solution.

$$Normality = \frac{Mass\ of\ the\ substance\ in\ one\ litre}{Gram\ equivalent\ mass\ of\ the\ substance}$$

**Example:** 1.0 N HCl solution contains 36.45 gm of HCl in one litre of d/w

### B. Molarity

It is a unit of concentration measuring the numbers of moles of a solute per litre of solution. Molarity is denoted with 'M'. 1M solution contains 1 mole of solute per litre of solution.

**Example:** 1.0 M HCl consists of 36.5gm HCl in 1000 ml of distilled water

### C. Molality

It is the number of moles of solute per kilogram of solvent. Solutions expressed in molal concentration are denoted with 'm'. 1 m solution contains 1 mole of solute per kilogram of solvent.

**Example:** 1.0 m $H_2SO_4$ solution contains 98gm of $H_2SO_4$ in 1000ml of water

### D. Formality

It is defined as the number of formula mass of any solute dissolved in one litre of solution. A solution is said to be one formal (1F) if it contains one formula mass of the substance in one litre of solution.

**Example:** 1.0 F solution of Oxalic acid ($H_2C_2O_4$, $2H_2O$) contains 126gm of oxalic acid in one litre of solutions

## Standard Curve

A standard curve also known as calibration curve, is a type of graph used as a quantitative research technique. Standard curves represent the relationship between two quantities. Multiple samples with known concentration are measured and plotted on the graph, which then allows similar concentrations to be determined for unknown samples by interpolation on the graph. The samples with known concentration are the standards, and the graph is the standard graph.

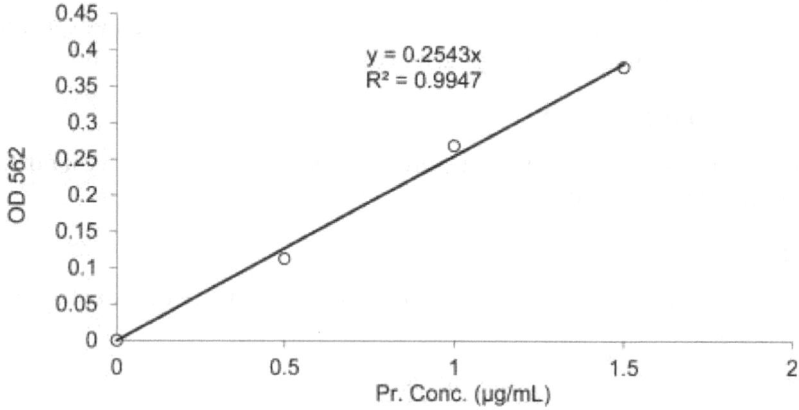


**Example:** A standard curve for protein concentration determination.

## Nomograph

- Nomographs are visual tools representing mathematical relationships.
- They use intersecting points on scales to determine variable values.
- Variables can be chosen arbitrarily for determination.
- Examples include calculating lift for hot-air balloons and boiling points under reduced pressure.
- Nomographs offer quick and reliable solutions, even without calculators.
- They are practical for field use due to their simplicity and effectiveness.

**Pressure-Temperature Nomograph**

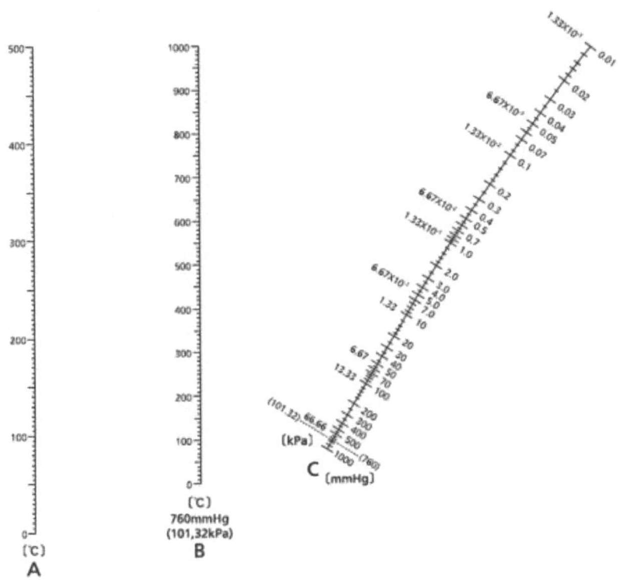


•How to calculate the bp under atmospheric pressure from bp under reduced pressure

① Connect a degree on the line C and its corresponding bp on the line A under reduced pressure using a straight line.

② An intersection found by step ① on the line B serves as an approximate bp in atmospheric pressure.

* This nomograph applies to nonassociated solvent.
Since the bp obtained from this nomograph is an approximate value, it is not an exact bp.

Chapter-2

# Chemistry of Water

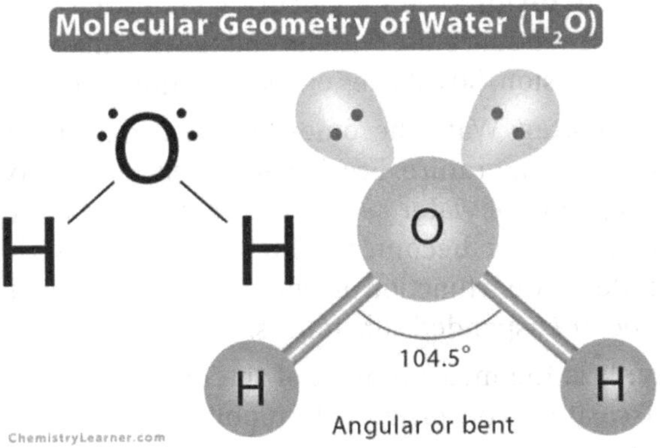


Water, chemically represented as $H_2O$, is a molecule composed of two hydrogen atoms bonded to one oxygen atom. It is a crucial substance for life on earth and has unique chemical properties due to its structure.

## Molecular Structure

- **Covalent Bonds**: The oxygen atom forms two covalent bonds with two separate hydrogen atoms. The oxygen hydrogen bonds are polar, meaning there is an uneven distribution of electrons, with the oxygen end being slightly negative and the hydrogen ends slightly positive.
- **Polarity:** The polarity of water results in a dipole moment, with the oxygen side having a partial negative charge (δ-) and the hydrogen side having a partial positive charge (δ+). This dipole moment gives water its unique properties.

- **Hydrogen Bonds:** The partial charges on water molecules allow for the formation of hydrogen bonds. The oxygen of one water molecule is attracted to the hydrogen of another. Hydrogen bonding is responsible for many of water's physical and chemical properties, including its high boiling and melting points.

## Physical Properties of Water

- **Specific heat:** The specific heat is the amount of heat per unit mass required to raise the temperature by one degree Celsius. Unit of specific heat is "Calorie". The calorie is defined as the amount of heat required to raise the temperature of one gram of water by one-degree Celsius (4.1868 Joule). Water has to absorb 4.184 Joules of heat for the temperature of one gram of water to increase 1 degree Celsius (°C). Specific heat of water is 1 Cal. In natural waters dissolved substances lower the specific heat.
- **Latent Heat of vaporization (LHV):** Vaporization is a process in which thermal energy make the water molecules loose from the liquid surface and into the gaseous state. Water has an exceptionally high latent heat of vaporization. LHV is the heat that required vaporizing 1.0 gm. of water at $100^0$C (540 Cal/gm).
- **Latent Heat of fusion:** The amount of heat required to change water from a solid state to a liquid state with no change in temperature is termed the latent heat of fusion. 80 calories of heat is required to melt 1g of ice at $0^0$C.
- **Density of water:** Water becomes less dense at high temperature and denser at low temperature. Fresh water reaches its maximum density at the freezing point of water i.e. $3.98^0$C. As it cools below $4^0$C, the density decreases i.e. water becomes lighter due to formation of crystalline structure. Density is a function of temperature, pressure and concentration of dissolved or suspended substances.
- **Viscosity:** It is the measurement of the internal molecular friction of a liquid. Viscosity is concerned with mobility and flow of water. Viscosity of water decreases with increasing temperature.
- **Adhesion:** Adhesion is the tendency to cling to the surface of materials by means of hydrogen bonds of water molecule and oxygen atoms of the other substances.
- **Cohesion:** Cohesion is the property of liquids which offers resistance to being pulled apart or to the formation of new surfaces.
- **Surface tension:** At the surface of a water mass, the force responsible for resistance is termed as surface tension. Surface tension could be defined as the property of the surface of a liquid that allows it to resist an external force, due to the cohesive nature of the water molecules. The surface tension of water is 72.5 dynes/cm at 25°C. It would take a force of 72.5 dynes to break a surface film of water 1 cm long. The surface tension of

water decreases significantly with temperature. Species associated with the surface film are known as neuston.

## Chemical Properties of Water

There are mainly 3 types of chemical reactions that could take place in water: Precipitation, Acid/base, and oxidation-reduction reaction.

- **Precipitation reaction:** Dissolved ions react with each other and form a solid compound or precipitate. A typical precipitation reaction that occurs in water is the formation of calcium carbonate solid when solution of calcium is mixed with solution of carbonate (Davis and Cornwell, 2012).

$$Ca^{2+}_{(aq)} + CO_3^{2-}{}_{(aq)} \qquad CaCO_{3(s)}$$

- **Neutralisation reaction:** A neutralization reaction is a reaction of an acid and a base that results in an ionic compound and possibly water. When a base is added to an acid solution, the acid is said to be neutralized. The ionic compound that is produced in a neutralization reaction is called a salt (Chang, 2010; Ebbing and Gammon, 2009). The net equation for most acid base reaction is:

$$H^+_{(aq)} + OH^-_{(aq)} \qquad H_2O_{(l)}$$

- **Oxidation-reduction reaction:** It is a reaction in which electrons are transferred between species or in which atoms change oxidation number. The substance that loses the electron is said to be oxidized and it is called a reducing agent. While the substance that accept the electron is said to be reduced and is called oxidizing agent. When one species release electron, another one should be available to accept the electron. Iron pipe corrosion is an example of oxidation–reduction reaction in which iron metal oxidizes and loses two electrons, while hydrogen ions accept those electrons and reduces to hydrogen gas (Chang, 2010; Davis and Cornwell, 2012; Ebbing and Gammon, 2009)

$$Fe \qquad Fe^{2+} + 2e^-$$

$$2H^+ + 2e^- \qquad H_{2\,(g)}$$

**Properties of pure water, freshwater and seawater**

| Properties | Seawater | Pure water | Fresh water |
|---|---|---|---|
| Specific Heat | Lower than pure water due to dissolved substances. 3.9 - 4.2 (J/g°C) | High, around 1 calorie per gram per degree Celsius. 4.18(J/g°C) | Similar to pure water, but may vary slightly depending on dissolved substances. 4.17 (J/g°C) |
| Latent Heat of Vaporization | Higher than pure water, typically around 540 Cal/gm. | High, around 540 Cal/gm. | Similar to pure water, around 540 Cal/gm. |
| Latent Heat of Fusion | Similar to pure water, around 80 calories per gram at 0°C. | 80 calories per gram at 0°C. | 80 calories per gram at 0°C. |

| Properties | Seawater | Pure water | Fresh water |
|---|---|---|---|
| Density | Higher than pure water due to dissolved salts. 1.025 - 1.029 (g/cm$^3$) | Maximum at 4°C, decreases above and below this temperature. 0.99997 (g/cm$^3$) | Similar to pure water, with maximum density at 4°C. 0.998 (g/cm$^3$) |
| Viscosity | Typically, higher than pure water due to dissolved substances. 1.1 - 1.3 (mPa.s) | Decreases with increasing temperature. 0.89 - 1.1 (mPa.s) | Similar to pure water, but may vary slightly depending on dissolved substances. 0.89 - 1.1 (mPa.s) |
| Adhesion | Sea water contains dissolved salts and ions which can lead to greater adhesion between water molecules and other substances. | Pure water also shows high adhesion due to hydrogen bonding between water molecules and surfaces. | Fresh water exhibits high adhesion due to hydrogen bonding between water molecules and surfaces. |
| Cohesion | Cohesion in sea water is influenced by dissolved salts and ions, leading to variable cohesion properties. | Pure water shows high cohesion due to the presence of hydrogen bonds between water molecules. | Fresh water exhibits high cohesion due to strong hydrogen bonding between water molecules. |
| Surface Tension | Generally higher than pure water, around 72.5 dynes/cm at 25°C. decreases with temperature. | Generally high, decreases with temperature. | Similar to pure water, but may vary slightly depending on dissolved substances. |
| Chemical Properties | Sea water contains dissolved salts, ions, and minerals which affect its chemical properties such as conductivity and pH. | Pure water is free from dissolved salts and ions, making it chemically pure with a neutral pH. | Fresh water is relatively pure compared to sea water, containing fewer dissolved substances. |

## Composition of Sea Water, Groundwater, and Surface Water

### Sea Water

Sea water, comprising the vast majority of the earth's water is saline and found in oceans and seas across the globe. Understanding its composition is crucial for marine science, environmental studies, and various industries.

#### Salinity and Major Ions

- Sea water is characterized by its high salinity, averaging about 35 parts per thousand (ppt). This salinity is primarily due to dissolved salts, with sodium chloride (NaCl) being the most abundant.
- Major ions in sea water include sodium ($Na^+$), chloride ($Cl^-$), magnesium ($Mg^{2+}$), sulphate ($SO_4^{2-}$), calcium ($Ca^{2+}$), and potassium ($K^+$).

### Trace Elements

- Sea water contains essential trace elements vital for marine life. These include iron (Fe), manganese (Mn), zinc (Zn), copper (Cu), and iodine (I).
- Trace elements play critical roles in biological processes, enzyme function, and overall ecosystem health within marine environments.

### Biological Diversity

The composition of sea water supports a rich and diverse array of marine life, from microscopic phytoplankton to large marine mammals.

## Groundwater

Groundwater, found beneath the earth's surface within aquifers, represents a vital freshwater resource essential for drinking, agriculture, and industry. Its composition is influenced by geological factors, recharge sources, and human activities.

### Minerals and Ions

- Groundwater often exhibits higher mineral content compared to surface water due to prolonged contact with subsurface rocks and minerals.
- Common ions found in groundwater include calcium ($Ca^{2+}$), magnesium ($Mg^{2+}$), sodium ($Na^{+}$), bicarbonate ($HCO_3^{-}$), sulphate ($SO_4^{2-}$), and chloride ($Cl^{-}$).

### Redox Conditions

- Redox conditions in groundwater can vary, influencing the mobility and speciation of certain elements like iron (Fe), manganese (Mn), and arsenic (As).

### Contamination Risks

- Groundwater is susceptible to contamination from various sources, including industrial activities, agricultural runoff, and improper waste disposal.
- Monitoring and managing groundwater quality are essential for protecting human health and safe guarding ecosystems dependent on groundwater resources.

## Surface Water

Surface water encompasses rivers, lakes, and streams, representing a vital component of the hydrological cycle. Its composition varies based on geographical location, climate, and anthropogenic influences.

### Physical Characteristics

- Surface water exists in a liquid state and is subject to variations in temperature, flow rates, and turbidity.

- Factors such as temperature, flow dynamics, and sediment load influence the physical properties of surface water bodies.

**Chemical Components**

- Surface water contains dissolved ions derived from geological weathering, atmospheric deposition, and human activities such as agriculture and industry.
- Common ions include calcium ($Ca^{2+}$), magnesium ($Mg^{2+}$), sodium ($Na^+$), potassium ($K^+$), bicarbonate ($HCO_3^-$), sulphate ($SO_4^{2-}$), and chloride ($Cl^-$).

**Ecological Significance**

- Surface water ecosystems support diverse biological communities, including fish, amphibians, aquatic plants, and microorganisms.
- Understanding surface water composition and quality is essential for maintaining healthy aquatic ecosystems and ensuring sustainable water resource management.

In conclusion, understanding the composition of sea water, groundwater, and surface water is fundamental for various scientific disciplines and practical applications, ranging from marine biology to water resource management.

Chapter-3

# Dissolved Gases

Dissolved gases are essential components of aquatic ecosystems, playing critical roles in oxygenation, carbon cycling, and overall water chemistry. Oxygen, carbon dioxide, nitrogen, and methane are among the key dissolved gases influencing biological processes and ecosystem dynamics. Understanding the dynamics of dissolved gases is fundamental for assessing water quality, ecosystem health, and the resilience of aquatic habitats to environmental changes.

## Dissolved Oxygen

Dissolved oxygen (DO) is a critical component of aquatic environments, essential for the survival of various organisms and indicative of water quality. It refers to the amount of oxygen gas that is dissolved in water, typically measured in milligrams per liter (mg/l) or parts per million (ppm). DO levels are influenced by various factors, including temperature, salinity, pressure, and biological activity.

Temperature plays a significant role in DO levels, with colder water holding more dissolved oxygen than warmer water. This is because the solubility of oxygen in water increases as temperature decreases. Additionally, DO solubility decreases as salinity increases, meaning freshwater typically holds more dissolved oxygen than seawater.

DO levels in aquatic environments can vary widely, ranging from supersaturated levels where more oxygen is present than can be held in solution, to hypoxic or anoxic conditions where oxygen levels are severely depleted. Dissolved oxygen is vital for the respiration of aquatic organisms such as fish, invertebrates, and bacteria. Insufficient DO can lead to stress, impaired growth, and even mortality in these organisms.

Monitoring dissolved oxygen levels is crucial for assessing water quality and ecosystem health. Low DO levels can be indicative of pollution, excessive nutrient input, or other environmental stressors. Conversely, high DO levels may indicate a healthy, well-oxygenated environment. Managing and maintaining adequate DO levels in aquatic ecosystems is essential for supporting diverse and thriving communities of organisms.

## Factors Influencing Oxygen Concentration

- **Temperature**: Oxygen solubility in water decreases as temperature increases. Colder water can hold more dissolved oxygen than warmer water. This is because molecules move more freely in warmer water, making it harder for oxygen molecules to remain dissolved.
- **Salinity**: The presence of dissolved salts affects oxygen solubility in water. Generally, freshwater holds more dissolved oxygen than saltwater. This is because salt molecules occupy space in the water, reducing the available space for oxygen molecules to dissolve.
- **Pressure**: Pressure influences the solubility of gases in water. At higher pressures, such as in deep ocean waters, oxygen can dissolve more readily. However, changes in pressure due to factors like altitude have minimal effects on oxygen solubility in surface waters.
- **Turbulence and Mixing**: Oxygen can be introduced into water through mixing and turbulence, such as from wave action or waterfalls. This helps aerate the water and increase oxygen levels. Conversely, stagnant or poorly mixed water bodies may have lower oxygen concentrations.
- **Photosynthesis and Respiration**: Aquatic plants and algae produce oxygen through photosynthesis during daylight hours, contributing to higher oxygen levels. Conversely, both plants and animals consume oxygen through respiration, especially at night or in areas with high biological activity, which can lower oxygen concentrations.
- **Pollution and Nutrient Loading**: Pollution from sources like agricultural runoff or wastewater discharge can lead to oxygen depletion in water bodies. Excess nutrients can promote algal blooms, which, when they die and decompose, consume large amounts of oxygen in a process called eutrophication.
- **Seasonal and Weather Patterns**: Oxygen concentrations can fluctuate seasonally and in response to weather patterns. For example, warm summer temperatures can lead to oxygen depletion in deeper waters through a process called stratification, where layers of water with different temperatures do not mix well.

## Dissolved Oxygen in Water Column

### A. Surface Zone (Epilimnion or Epipelagic Zone)

- This is the uppermost layer of the water column, generally exposed to sunlight.
- Dissolved oxygen levels are typically highest here due to the diffusion of atmospheric oxygen and photosynthesis by aquatic plants.
- However, oxygen levels can fluctuate depending on factors such as temperature and biological activity.

### B. Middle Zone (Metalimnion or Mesopelagic Zone)

- This is the middle layer of the water column, characterized by a rapid decrease in temperature and light penetration.
- Dissolved oxygen levels may vary depending on mixing patterns, biological activity, and oxygen consumption by organisms.

### C. Deep Zone (Hypolimnion or Bathypelagic Zone)

- This is the deepest layer of the water column, often colder and darker than the surface layers.
- Dissolved oxygen levels can be lower in this zone due to limited light penetration, reduced mixing with surface waters, and increased decomposition of organic matter.
- Oxygen levels may decrease over time as organic material sinks from the surface and undergoes microbial decomposition.

### D. Bottom Zone (Benthic Zone)

- This refers to the sediment or substrate at the bottom of a body of water.
- Dissolved oxygen levels can vary greatly depending on factors such as sediment composition, organic matter content, and microbial activity.
- In some cases, oxygen levels may be depleted near the sediment surface due to microbial respiration and the breakdown of organic matter.

The distribution of dissolved oxygen within these zones can also be influenced by factors such as water circulation patterns, nutrient inputs, pollution, and seasonal changes. Monitoring dissolved oxygen levels and understanding its zonation is essential for assessing water quality, ecosystem health, and the habitat suitability for aquatic organisms.

## Carbon Dioxide

- **Environmental Balance**: Carbon dioxide helps keep the environment balanced by controlling how acidic or alkaline it is. This balance is important for the health of ecosystems.

- **Plant Growth**: Plants use carbon dioxide to grow and make their food through a process called photosynthesis. It's like their version of eating. Without enough carbon dioxide, plants can't grow properly.
- **Seed Germination**: Some plants need carbon dioxide to start growing from their seeds. It's like a kickstart for their growth process.
- **Chemical Reactivity**: Carbon dioxide mixes easily with water and other elements. It reacts with them to create different kinds of compounds. This is important for many chemical processes in nature.
- **Ocean Chemistry**: In the ocean, carbon dioxide combines with other chemicals like bicarbonates and carbonates. This affects the chemistry of the water and helps regulate its acidity. It's crucial for marine life.
- **Global Carbon Cycle**: Carbon dioxide is part of a cycle that moves carbon through the environment. It's released into the atmosphere through activities like burning fossil fuels. Then, plants and oceans absorb it, helping to balance its levels.

## Other Gases

### Methane ($CH_4$)

- Methane, also known as marsh gas, is an organic gas found in alkaline lakes, ponds, and swamps, especially during the summer season.
- It is produced chemically due to the decomposition of organic substances under anaerobic conditions in aquatic environments.

### Nitrogen ($N_2$) and Argon (Ar)

- Nitrogen and argon are also present in natural waters, primarily as dissolved gases.
- Nitrogen is relatively insoluble in water but can still be found in trace amounts.
- Argon, being inert, remains dissolved in water without reacting chemically.

### Hydrogen Sulphide ($H_2S$)

- Hydrogen sulphide can be present in natural waters, particularly in anaerobic conditions and environments with organic decomposition.

**Factors affecting natural waters**

| Factors Affecting Natural Water | Description | Example |
|---|---|---|
| Geological Composition | The type of rocks and soils in an area can influence water chemistry. Different rock types release different minerals into water, affecting its composition. | Limestone areas can increase calcium and bicarbonate levels in groundwater due to the dissolution of calcium carbonate. |

| Factors Affecting Natural Water | Description | Example |
|---|---|---|
| Climate and Weather | Precipitation, temperature, and evaporation rates influence water levels, flow rates, and temperature in water bodies. | Climate change alters precipitation patterns, affecting water availability and quality. |
| Land Use and Human Activities | Human activities such as agriculture, urbanization, and industry introduce pollutants like nutrients, pesticides, and heavy metals into water bodies. | Fertilizers from farms can lead to nutrient runoff into rivers, causing algal blooms and affecting aquatic ecosystems. |
| Hydrology and Hydrogeology | Water movement through the hydrological cycle, including infiltration, runoff, and groundwater flow, affects water availability and quality. | Groundwater contamination from industrial sites can spread through aquifers, affecting drinking water sources. |
| Biological Interactions | Aquatic plants, algae, and microorganisms impact water quality through processes like photosynthesis and nutrient cycling. | Algal blooms can occur when excess nutrients are present, leading to oxygen depletion and harming fish and other aquatic life. |
| Topography and Drainage Patterns | The shape of the land and drainage patterns determine how water flows across the landscape, affecting erosion rates and sediment transport. | Steep slopes can increase erosion, leading to sedimentation in rivers and lakes. |
| Human Management Practices | Water management strategies like dam construction and irrigation can alter natural water systems, affecting flow regimes and ecosystem health. | Dams can regulate river flow but may also disrupt downstream habitats and fish migration. |
| Natural Disturbances | Natural events such as floods, droughts, and volcanic eruptions can impact water bodies by altering flow patterns and introducing materials like ash and debris. | Floods can carry sediment and pollutants downstream, affecting water quality and habitat integrity. |

Chapter-4

# Acids, Bases and Salts

## Acids and Bases

For thousands of years, people have known that vinegar, lemon juice, amla, tamarind, and many other food items taste sour. However, it was only a few hundred years ago that it was proposed that these things taste sour because they contain 'acids'. The term 'acid' comes from the Latin term 'accre', which means sour. It was first used in the seventeenth century by Robert Boyle to label substances as acids and bases according to the following characteristics:

| Acids | Bases |
|---|---|
| 1. Taste sour | 1. Taste bitter |
| 2. Corrosive to metals | 2. Feel slippery or soapy |
| 3. Change blue litmus into red | 3. Change red litmus into blue |
| 4. Become less acidic on mixing with bases | 4. Become less basic on mixing with acids |

## Acids

An acid is a substance which furnishes hydrogen ions ($H^+$) when dissolved in water. For example, in its aqueous solution hydrochloric $HCl_{(aq)}$ dissociates as:

$$HCl_{(aq)} \longrightarrow H^+_{(aq)} + Cl^-_{(aq)}.$$

Some examples of acids are:

- Hydrochloric acid (HCl) in gastric juice
- Carbonic acid ($H_2CO_3$) in soft drinks

- Ascorbic acid (vitamin C) in lemon and many fruits
- Citric acid in oranges and lemons
- Acetic acid in vinegar
- Tannic acid in tea, Nitric acid ($HNO_3$) used in laboratories
- Sulphuric acid ($H_2SO_4$) used in laboratories

## Bases

A base is a substance which furnishes hydroxide ions ($OH^-$) when dissolved in water.

For example, sodium hydroxide $NaOH_{(aq)}$ in its aqueous solutions, dissociates as:

$$NaOH_{(aq)} \longrightarrow Na^+_{(aq)} + OH^-_{(aq)}$$

The term 'alkali' is often used for water soluble bases. Some examples of bases are:

- Sodium hydroxide (NaOH) or caustic soda used in washing soaps.
- Potassium hydroxide (KOH) or potash used in bathing soaps.
- Calcium hydroxide ($Ca(OH)_2$) or lime water used in white wash.
- Magnesium hydroxide ($Mg(OH)_2$) or milk of magnesia used to control acidity.
- Ammonium hydroxide ($NH_4OH$) used in hair dyes.

## Types of Acids

| Acid Type | Definition | Examples |
|---|---|---|
| Strong Acids | Acids that completely dissociate in water, yielding a high concentration of $H^+$ ions. | Hydrochloric acid (HCl) Sulfuric acid ($H_2SO_4$) Nitric acid ($HNO_3$) Hydrobromic acid (HBr) Hydroiodic acid (HI) Perchloric acid ($HClO_4$) Chloric acid ($HClO_3$) |
| Weak Acids | Acids that only partially dissociate in water, yielding a low concentration of $H^+$ ions. | Acetic acid ($CH_3COOH$) Citric acid (found in citrus fruits) Carbonic acid ($H_2CO_3$) Formic acid (HCOOH) |
| Concentrated Acids | Solutions containing a high amount of acid molecules or ions in a given volume of solvent. | Concentrated sulfuric acid ($H_2SO_4$) Concentrated hydrochloric acid (HCl) Concentrated nitric acid ($HNO_3$) |
| Dilute Acids | Solutions containing a low amount of acid molecules or ions in a given volume of solvent. | Dilute sulfuric acid ($H_2SO_4$) Dilute hydrochloric acid (HCl) Dilute nitric acid ($HNO_3$) |

## Properties of Acids

The important properties of acids are as follows:

i. Acids have a sour taste.

ii. Acids turn blue litmus to red.

iii. Acids solution conducts electricity due to the presence of hydrogen ions.

iv. Acids have a corrosive nature.

v. Acids react with metals to form hydrogen gas.

Metal + Acid → Salt+ Hydrogen gas

For example, when dilute sulphuric acid reacts with zinc metal, then zinc sulphate is formed with the liberation of hydrogen gas.

$$Zn + H_2SO_4 \rightarrow ZnSO_4 + H_2$$

Zinc + Sulphuric acid → Zinc sulphate + Hydrogen gas

vi. Strong acids react with metal carbonate and metal hydrogen carbonate and form carbon dioxide gas.

Metal carbonate + Acid → Salt + Water + Carbonate dioxide

For example, when dilute hydrochloric acid reacts with sodium carbonate, then sodium chloride, water and carbon dioxide gas are formed:

$$Na_2CO_3 + 2HCl \rightarrow 2NaCl + CO_2 + H_2O$$

Sodium carbonate + Hydrochloric acid → Sodium chloride + Carbon dioxide + Water

vii. Acids react with a base to form salt and water. The reaction between an acid and a base to form salt is known as a neutralisation reaction.

Acid + Base → Salt + Water

For example, when hydrochloric acid reacts with sodium hydroxide solution, a neutralization reaction takes place and forms sodium chloride and water

viii. Acids react with metal oxides and form salt and water.

For example, when hydrochloric acid reacts with copper (II) oxide to form copper (II) chloride and water.

$$CuO + 2HCl \rightarrow CuCl_2 + H_2O$$

Copper (II)oxide + Hydrochloric acid → Copper (II)chloride + Water

## Uses of Acids in Industry

- Sulphuric acid is used in the manufacture of fertilisers, paints, chemicals, plastics, detergents, etc.
- Nitric acid is used in the manufacture of fertilisers, explosives, dyes and plastics.

- Hydrochloric acid is used to remove oxide film from steel objects as well as scale deposits from within boilers. It's also used in dyestuffs, textiles, food, and leather.

## Bases

Bases are those chemical substances that have a bitter taste. Bases are the chemical opposite to acids. A base is a substance that can neutralize an acid. All the bases change the colour of red litmus to blue. The pH of the basic solution is above 7.7. Some common examples of bases are sodium hydroxide, magnesium hydroxide, calcium hydroxide, magnesium hydroxide, etc.

A base that is soluble in water is called alkali. Sodium hydroxide, potassium hydroxide, calcium hydroxide are some examples of alkalis. Bases have hydroxide ions in common. Thus, a base is a substance that dissolves in water to produce hydroxide ions ($OH^-$) in the solution.

### Types of Bases

Bases are classified into two types. They are;

**Strong Bases:** A base that completely ionises in water and thus produces a large amount of hydroxide ions ($OH^-$) is called a strong base (or a strong alkali). Eg: Sodium hydroxide (NaOH)

**Weak Bases:** A base that is partially ionised in water and thus produces a small amount of hydroxide ions ($OH^-$ions) is called a weak base. Ammonium hydroxide, calcium hydroxide and magnesium hydroxide are weak bases.

### Properties of Bases

The important properties of bases are as follows:

- Bases have a bitter taste
- Bases feel soapy to touch
- Bases turn red litmus to blue
- Bases conduct electricity due to the presence of hydroxide ions.
- Bases react with some metals to form salt and hydrogen gas. For instance, sodium and hydrogen gas are formed if the solution for sodium hydroxide is heated by zinc.
- Bases react to salt and water with acids. This reaction is called a reaction of neutralisation. For instance, sodium chloride and water are formed when hydrochloric acid reacts. This is an example of a reaction of neutralisation.

$$NaOH + HCl \rightarrow NaCl + H_2O$$

- Bases react with non-metal oxides to form salt and water.

Non metal oxide + Base → Salt + Water

For example, calcium hydroxide reacts with carbon dioxide and produces calcium carbonate and water.

$$Ca(OH)_2 + CO_2 \rightarrow CaCO_3 + H_2O$$

Calcium hydroxide + Carbon dioxide → Calcium carbonate + Water

## Uses of Bases

The following are some of the most important applications of bases:

- Sodium hydroxide is used in the production of soap, paper, and rayon, a synthetic fibre.
- Calcium hydroxide (also known as slaked lime) is used in the production of bleaching powder.
- Magnesium hydroxide is used as an antacid to neutralise excess stomach acid and cure indigestion.
- Sodium carbonate is used to soften hard water and as a washing soda.
- Sodium hydrogen carbonate is used in baking soda, baking powders, as an antacid to treat indigestion, and in soda-acid fire extinguishers.

# Salts

Salt is an acid-formed compound that substitutes for acid hydrogen with a metal. When acids react with the foundations, salts are formed. The pH of the salt solution is generally 7.7. The most well-known salt is sodium chloride (NaCl) also known as common salt. Other salts are calcium chloride, magnesium chloride, magnesium chloride, zinc chloride, aluminium sulphate, zinc carbonate, sodium acetate, etc.

## Properties of Salts

- Salts are mostly solids.
- Salts are ionic compounds. Hence, they have high melting points and boiling points.
- Salts are usually soluble in water
- Just like acids and bases, the salt solution also conducts electricity due to the presence of ions in them.

## The pH of the Salt Solution

The aqueous solutions of many salts are neutral (having a pH of 7.7), but when dissolved in water, some salts produce acidic or basic solutions. The acidic nature and basic nature of some salt solutions can be explained based on the hydrolysis of salts.

## Types of Salts

Salts are of three types. They are as follows:

- The salts of strong acids and strong bases give a neutral solution.
- The salts of strong acids and weak bases give acidic solutions.

- The salts of weak acids and strong bases give basic solutions.

## Uses of Salts

- Common salt is used as a raw material for making a large number of useful chemicals in the industry such as sodium hydroxide, sodium carbonate, sodium hydrogen carbonate, hydrochloric acid, chlorine, hydrogen and sodium metal.
- Common salt is used in cooking food.
- Common salt is used as a preservative in pickles, and in currying meat and fish.
- Common salt is used in the manufacture of soap.
- Washing soda is a transparent crystalline solid.
- Sodium carbonate is used in the manufacture of sodium compounds such as borax.
- Sodium hydrogen carbonate is used as an antacid in medicine to remove the acidity of the stomach.

## Ionization of Water

The ionization of water refers to the process by which water molecules dissociate into ions. In pure water, a small fraction of water molecules naturally undergoes this process. Water can ionize into a positively charged hydrogen ion ($H^+$) and a negatively charged hydroxide ion ($OH^-$). The chemical equation for this process is represented as:

$$H_2O \rightleftharpoons H^+ + OH^-$$

This equilibrium reaction indicates that water can exist in a dynamic equilibrium where it can dissociate into ions and recombine. The concentration of $H^+$ and $OH^-$ ions in pure water at 25 degrees Celsius is $1 \times 10^{-7}$ moles per liter each. This concentration is often referred to as the "neutral" pH of water.

The ionization of water is an essential concept in understanding the properties of aqueous solutions, acids, and bases. It also plays a crucial role in the pH scale, which measures the acidity or alkalinity of a solution. In neutral water, the concentration of $H^+$ ions equals $OH^-$ ions, resulting in a pH of 7.

## Hydrogen Ions ($H^+$)

**Formation and Structure:** Hydrogen ions are formed when a hydrogen atom loses its electron, leaving behind a lone proton. Thus, they consist of a single positively charged proton. Due to their small size and high charge density, hydrogen ions exhibit strong electrostatic interactions with other molecules and ions.

**Behaviour in aqueous solutions:** In water, hydrogen ions interact with water molecules through protonation, leading to the formation of hydronium ions ($H_3O^+$).

The equilibrium between hydrogen ions and hydronium ions is described by the equation:

$$H_2O + H^+ \rightleftharpoons H_3O^+$$

**Acid-Base Chemistry:** Hydrogen ions are central to acid-base reactions. Acids donate hydrogen ions to solutions, increasing the concentration of $H^+$ ions, while bases either accept hydrogen ions or donate hydroxide ions ($OH^-$) that can react with $H^+$ ions.

## Modern Concept of pH

The modern concept of pH is a measure of the acidity or alkalinity of a solution. It's based on the concentration of hydrogen ions ($H^+$) present in the solution. The pH scale ranges from 0 to 14, with 7 considered neutral. Solutions with a pH less than 7 are acidic, while those with a pH greater than 7 are alkaline or basic.

The pH scale is logarithmic, meaning that each whole number change on the scale represents a tenfold difference in acidity or alkalinity. For example, a solution with a pH of 5 is ten times more acidic than a solution with a pH of 6.

The concept of pH was introduced by Danish chemist Søren Peder Lauritz Sørensen in 1909 as a way to simplify the expression of hydrogen ion concentration. It has since become a fundamental concept in chemistry, biology, environmental science, and many other fields. In addition to measuring acidity and alkalinity, pH is also important because it affects chemical reactions, biological processes, and the behaviour of substances in solution. For example, pH influences the solubility of minerals, the effectiveness of enzymes, and the growth of organisms in aquatic ecosystems.

Overall, the modern concept of pH provides a quantitative way to understand and describe the acidity or alkalinity of solutions, making it a valuable tool in scientific research and practical applications.

## Buffers

Buffers are solutions that resist changes in pH upon addition of acid or base. They consist of a weak acid and its conjugate base or a weak base and its conjugate acid. Buffers work by undergoing reversible reactions that consume added $H^+$ or $OH^-$ ions, thereby maintaining the pH within a relatively narrow range.

**Buffer Capacity:** The effectiveness of a buffer is determined by its buffer capacity, which is the amount of acid or base that can be added to the buffer without causing a significant change in pH. Buffers with higher concentrations of buffering components have higher buffer capacities and can resist larger changes in pH.

**Applications:** Buffers are widely used in various laboratory, industrial, and biological applications. They are essential for maintaining stable pH conditions in biochemical reactions, cell culture media, pharmaceutical formulations, and chemical synthesis processes.

**Examples:** Common examples of buffers include the acetic acid/acetate buffer ($CH_3COOH/CH_3COO^-$), phosphate buffer ($H_2PO_4^-/HPO_4^{2-}$), and Tris buffer (tris(hydroxymethyl)aminomethane). These buffers are used in different pH ranges and have specific applications based on their buffering capacity and compatibility with reaction conditions.

Chapter-5

# Water Analysis

## Collection, Preservation and Analysis of Water Samples

Collection of water sample is an integral part of the total analytical procedure. The amount of water sample collected from a water body should be a true representative of the water body under study. Water samples should always be collected keeping in mind the following points:

- To collect representative sample in suitable volume for analysis
- To maintain water properties original to the sample
- To keep the concentration of all components as in the original sample
- To transport sample with no significant changes in the composition of water

The frequency of sampling varies with the objective of investigation. However, for routine limnological work samples are collected fortnightly or at monthly intervals. For investigating public health related problems, more frequent sampling should be attempted.

For collecting surface water samples, a simple bucket or beaker may be used. However, for drawing sub-surface water sample, suitable equipment such as Vandorn sampler, Kammerer sampler or a Reversing bottle could be used.

### Vandorn Sampler

The Vondorn sampler, commonly known as a Van Dorn bottle, is a cylindrical device designed for collecting water samples from specific depths in aquatic environments. Its mechanism involves lowering the sampler into the water until it reaches the desired depth while keeping the valve closed to prevent water from entering prematurely. Upon reaching the target depth, a triggering mechanism releases the valve, allowing water to rush into the container and fill it with a sample from the specified depth. Once filled, the valve is closed again, sealing the sample inside the container. The sampler is then raised back to the surface using a rope or cable, and the collected sample can be retrieved for further analysis. This sampling method is widely used in environmental monitoring, oceanography, and limnology to study water quality, temperature gradients, and the distribution of organisms in aquatic ecosystems.

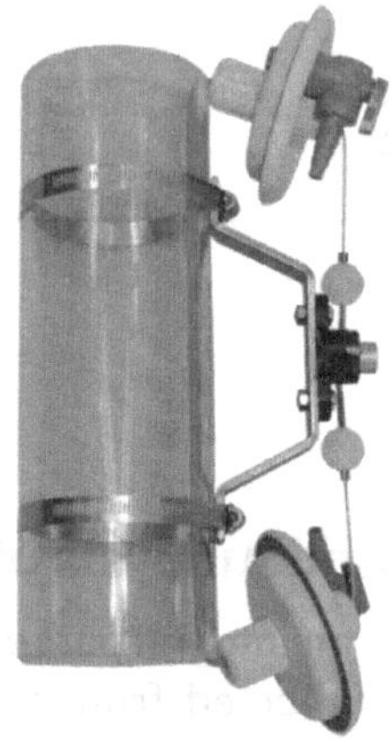

## Kemmerer Sampler

The Kemmerer sampler, devised by H.C. Kemmerer, is an essential tool for collecting water samples from distinct depths in bodies of water like lakes and rivers. Comprising a weighted cylindrical tube with spring-loaded chambers, it is deployed into the water, sinking vertically to the desired depth. Once positioned, the sampler triggers the release of spring-loaded chambers at pre-determined intervals, each collecting a discrete water sample from various depths simultaneously. These samples are then sealed within the chambers for retrieval. Upon resurfacing, the collected water samples can be analyzed to assess parameters such as temperature, dissolved oxygen levels, and pollutant concentrations, offering valuable insights into the vertical water profile and spatial distribution of water quality constituents within the aquatic ecosystem.

## Quantity of Sample

Generally, 2 liters or more volume of sample are considered to be sufficient for most of the routine parameters including pigments, TSS, TDS, metals etc. When collecting samples directly from a river, stream, lake, reservoir, spring or shallow well, the aim must be to obtain a representative sample. It is thus undesirable to take samples too close to the bank, far from the point of draw-off or near the bottom.

## Sample Containers and Labels

Normally, plastic or glass containers are used for collecting water samples. However, each has its own drawbacks. For example, silica and sodium may get leached from glass but not from plastic. Similarly, trace level metals may be adsorbed into the walls of glass containers. For samples having volatile organic, glass bottles are preferred. Each sample should be labelled bearing details of sample or appropriate code. A field log book should invariably be maintained for recording details of samples and analysis thereof. GPS readings may be incorporated for identification of samples.

## Preservation of Samples

Some water quality parameters get altered if the sample is not analysed quickly. Therefore, it is desirable to make at the site analysis of a few chemical parameters such as dissolved oxygen, carbon dioxide, alkalinity, pH and temperature.

For rest of the analyses, samples could be preserved or stored for a specific time for further analysis in the laboratory.

# Measurement of Physico-Chemical Parameters of Water

## Water Temperature

For determination of surface and subsurface water temperature of a water body generally 2 types of thermometers are used. For measuring surface water temperature, common centigrade thermometer which can measure from 0 to 100°C is generally used. For measurement of subsurface water temperature - reversing thermometer is widely used. For determining the temperature, dip the thermometer directly into the water, keep it for about one minute or till the reading become constant and note down the temperature immediately. Now a days, electronic thermometers are also available for measurement of water temperature.

## Transparency/Turbidity

Turbidity refers to an optical property of water that causes light to be scattered or absorbed rather than transmitted through the water in a straight line. Turbidity is caused by suspended material (such as soil particles, plankton, and organic detritus) and soluble coloured organic compounds. Turbidity caused by plankton generally is desirable in fish ponds.

## Secchi Disc Method

The Secchi Disc is a round metal plate with black and white sections, attached to a rope for lowering into water. It sinks due to a weight and is painted black on the bottom to minimize reflections. To use it, you lower it into the water and note when it disappears and reappears, averaging these depths for visibility. It's used between 9 am. and 3 pm. in sunny conditions from the sunny side of a boat to compare different waters or assess changes over time. However, it's not precise as factors like plankton and environmental conditions affect how light travels in water, making it a rough estimate of visibility.

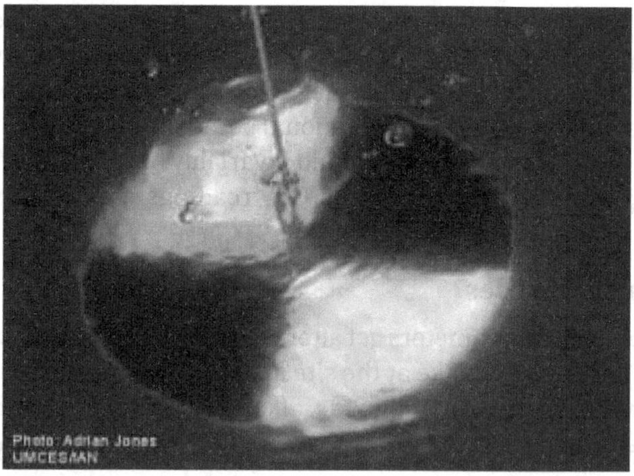


The transparency as a measure of light extinction coefficient (K) is calculated using the following formula of Poole and Atkins:

$$K = \frac{1.7}{d}$$

Where 'd' is the depth (m) of secchi disc visibility. For turbid waters, however a value of 1.44 is considered instead of 1.7.

For offshore waters

$$K = \frac{1.9}{d}$$

For shallow coastal waters

$$K = \frac{1.5}{d}$$

## Water pH

At field level approximate idea of pH of a water sample can be obtained by using simple pH paper or Litmus paper. pH of a water sample can also be measured colorimetrically and potentiometrically.

### a. pH Paper

- pH paper strips are easily available in market and are very handy for getting an approximate idea of pH value of a water sample.
- The steps involved in determination of pH of a water sample using pH paper are as follows:
- Take a single pH strip from the pH book.
- Dip the pH strip in your water sample and take out quickly.
- Allow the strip to dry for a minute or so.
- Compare the colour developed in the pH strip with the standard colours of pH books. Note down the corresponding pH of the matching standard colour.

### b. pH Meter

The method is more accurate; requires less time; widely used. pH is recorded by immersing the standardized pH meter by following guidelines of the manufacture. pH meter is calibrated with pH values of 4.0, 7.0 and 9.2 standard pH solutions prepared by using the tablets of each pH. Now a days, pH pens are also available which are very handy to use and gives accurate result.

## Electrical Conductivity

Electrical Conductivity indicates the total concentration of ionized constituents in water. Closely related to the amount of total dissolved solids. EC - used as an index of salt content of water. EC is reciprocal of electrical resistance. In natural water EC values normally ranges from 20 to 1500 mmhos /cm. EC = I/2 R, where R is resistance.

Electrical conductivity is measured using an electrical conductivity bridge known as Electrical conductivity meter/Whatman conductivity bridge. The conductance of the water sample is measured by immersing the cell of the conductivity meter and recorded the conductance value. By multiplying this value by cell constant value gives EC of the water sample

**Calculation:** EC (mmhos/cm) = measured conductivity X cell constant

## Salinity/Chlorinity

The halogen compounds present in seawater (chloride, bromide and iodide) readily react with silver to give insoluble silver halides. Fluoride (a halogen ion) does not precipitate with silver. Hence, total quantity of silver required to react with chloride, bromide and iodide is a true measure of chlorinity of seawater. In this method potassium chromate is used as the indicator. The main principle of this titrimetric method is that silver will not react with chromate till all the halogens (other than fluoride) present in seawater are converted into silver halides. Since silver chromate has a characteristic brownish red colour, the end point can be precisely noted.

### Reagents

- Silver nitrate solution
- Potassium chromate solution

### Procedure

100 ml water sample is taken in a conical flask and 1 ml of chromate solution is added to it. The entire sample now turns yellow. Slowly the silver nitrate already taken in the burette is added. The titration is stopped when the yellow colour vanishes with the formation of a brownish red colour. Values are then tabulated. The titration is repeated with fresh sample till concordant values are obtained.

### Calculation

ml of silver nitrate used (at the time of end point) for the titration of 10 ml = Salinity of the sample. Salinity of the seawater sample may be expressed by giving the unit ppt (parts per thousand). For example, if 35 ml of silver nitrate is required for the titration, the salinity of the water sample will be 35 ppt.

## Solids

### 1. Total Solids (TS)

- Represents all the solid materials present in a water sample, including dissolved and particulate organic and inorganic matter.
- Determined by evaporating a water sample and weighing the residue left behind.

### 2. Total Volatile Solids (TVS)

- Estimates the amount of dissolved and particulate organic matter.
- Measured by igniting the residue obtained from total solids and determining the weight loss.

### 3. Total Dissolved Solids (TDS)

- Estimates the amount of dissolved organic and inorganic matter.
- Determined by filtering the water sample to remove particulate matter, evaporating the filtrate, and weighing the residue.

### 4. Total Volatile Dissolved Solids (TVDS)

- Estimates the amount of dissolved organic matter.
- Obtained by igniting the residue obtained from total dissolved solids and determining the weight loss.

## Procedure

### Total solids (TS)

A 100 ml of water sample is taken into a well cleaned and pre-weighed dish. This is then evaporated in an oven at 103°C for 1 hr. Afterwards, the dish and residue are collected in a desiccator and are weighed subsequently

$$Total\ solids\ (mg/l) = \frac{R - D\ x\ 1000}{Sample\ Volume\ (ml)}$$

Where,

R = weight (g) of the evaporating dish and residue

D = weight (g) of the evaporating dish

### Total volatile solids (TVS)

The aforesaid dish containing the residue is then placed in a muffle furnace which is maintained at 550°C for about 30 minutes. The dish and residue are cooled in a desiccator before weighing them. The amount of total volatile solids is calculated as follows

$$Total\ volatile\ solids\ (mg/l) = \frac{R - A\ x\ 1000}{Sample\ Volume\ (ml)}$$

Where,

R= weight (g) of the evaporating dish and residue before ignition

A= weight (g) of the evaporating dish and residue after ignition

### Total Dissolved Solids (TDS)

125 to 150 ml of water sample is filtered through a glass fibre filter placed in a funnel. Then 100 ml of the filtrate is transferred to a pre-weighed dish. The contents are evaporated as for total solids. The dish and residue are cooled in a desiccator before weighing.

$$Total\ dissolved\ solids\ (mg/l)\frac{R - D\ x\ 1000}{Sample\ Volume\ (ml)}$$

Where,

R = the weight (g) of the evaporating dish and residue

D = the weight (g) of the evaporating dish

### Total volatile dissolved solids (TVDS)

The evaporating dish and residue obtained from the analysis of the total dissolved solids are placed in a muffle furnace maintained at 550$^0$C for about 30 minutes. The dish and residue are cooled in a desiccator before weighing.

$$Total\ volatile\ dissolved\ solids\ (mg/l) = \frac{R - A\ x\ 1000}{Sample\ Volume\ (ml)}$$

Where,

R = the weight (g) of the evaporating residue before ignition

A = the weight (g) of the evaporating residue after ignition

## Dissolved Oxygen

Dissolved oxygen is a vital parameter in the metabolic activity of flora and fauna of aquatic ecosystem. Main source of dissolved oxygen in natural waters are mainly attributed from photosynthetic and respiratory activities of biota and diffusion gradient at the air water interface and distribution by wind driven mixing. It can be measured by a volumetric method known as "Winkler method".

### Principle

Manganous sulphate reacts with potassium or sodium hydroxide to give a white precipitate of manganous hydroxide. In the presence of oxygen, brown manganic basic oxide is formed,

$Mn^{+2}\ SO_4 + 2KOH \rightarrow Mn^{+2}\ (OH)_2 + K_2SO_4$

$2Mn^{+2}\ (OH)_2 + O_2 \rightarrow 2Mn^{+4}\ O\ (OH)_2 - O_2$

Addition of sulphuric acid dissolves the brown manganic oxide yielding manganic sulphate which reacts instantly with iodide to yield iodine.

$Mn^{+4}\ O\ (OH)_2 + 2H_2SO_4 \rightarrow Mn^{+4}\ (SO_4)_2 + 3H_2O$

$Mn^{+4}\ (SO_4)_2 + 2KI \rightarrow Mn^{+2}\ SO_4 + K_2SO_4 + I_2$

In effect, oxygen oxidizes $Mn^{+2}$ to $Mn^{+4}$ and $Mn^{+4}$ oxidizes $I^-$ to $I_2$. Iodine is then determined titrimetrically with starch as an end point indicator.

## Reagents

- ☆ Manganese Sulphate (Winkler A)
- ☆ Alkaline iodide (Winkler B)
- ☆ Sulphuric Acid (50% v/v)
- ☆ Starch indicator (1%)
- ☆ Sodium thiosulphate (0.01 N)

## Procedure

When transferring the water sample, the sampler dispensing tube must be placed at the bottom of the sample bottles (BOD bottles) of 125 ml capacity and held below the liquid surface while filling. Liquid should be allowed to overflow the bottle by at least one bottle volume in order to avoid the entrapping of air bubbles. Immediately 1ml of manganous sulphate and 1 ml of alkaline iodide are added one after another. These reagents should be added well below the liquid surface with the sample bottles open for a very short time. The bottle is restoppered without entrapping any air bubbles and carefully tilted, when a precipitate forms and settles at the bottom. At the time of titration, the stopper of the bottle is removed and 1 ml of sulphuric acid (50% v/v) is added. The bottle is again tilted carefully, when the precipitate dissolves and a clear straw yellow solution results. During titration 100 ml of solution is removed from the bottle and is transferred to a 250ml conical flask. This solution is then titrated against 0.01 N sodium thiosulphate solution until a strong pale-yellow colour is obtained. At this stage, 3 or 4 drops of 1 percent starch indicator is added and the titration is continued until the disappearance of the blue colour. Water samples collected for the analysis of dissolved oxygen should not be stored for more than eight hours.

## Calculation

$$Dissolved\ Oxygen\ (mg/l) = \frac{CF \times N \times E \times 1000 \times Vt}{Vs}$$

**Where,**

CF = Correction factor value

$$Correction\ factor\ value = \frac{Total\ volume\ of\ the\ stoppered\ bottle}{Total\ volume\ of\ stoppered\ bottle\ -\ 2\ .}$$

| | | |
|---|---|---|
| N | = | Normality of thiosulphate (0.01 N) |
| E | = | Equivalent weight of O2 |
| 1000 | = | To convert the value to 1000 ml (or) 1 litre |
| Vt | = | Titre value |
| Vs | = | Volume of Sample taken for titration |

If the oxygen content is to be calculated in ml/l, then the above result should be multiplied by a factor value (0.698). In case of brackish water and seawater samples, it is customary to express the value of oxygen content in ml/l and in freshwater samples, mg/l.

## Free $CO_2$

Carbon dioxide is present in atmosphere in very small quantity. For this reason, in spite of its high solubility in water, its concentration in most water-bodies is low. It occurs in water in three closely related forms: a) free $CO_2$, b) bicarbonate ion ($HCO_3$), and c) carbonate ion ($CO_3^{2-}$). Each form amount present depends on the pH of water. For example, in neutral or acidic waters, high concentration of free carbon dioxide, i.e. in toxic form is frequently found. Most natural waters contain free carbon dioxide. The equilibrium concentration of $CO_2$ in water is about 0.7 mg/l. However, many waters used for fish culture and most ground waters are supersaturated with carbon dioxide. Essentially, no free carbon dioxide is present in water of pH 8.3 or above. Free $CO_2$ can be estimated by a volumetric method.

## Principle

Carbon dioxide reacts with NaOH as follows

$$2NaOH + CO_2 \; Na_2CO_3 + H_2O$$

$$Na_2CO_3 + H_2O + CO_2 \; 2NaHCO_3$$

These two reactions indicate the two possible methods for determining carbon dioxide in water, titration with either standard NaOH or with standard $Na_2CO_3$ to the phenolphthalein end point. For titration with NaOH, 0.0227 N solution is required. For titration with $Na_2CO_3$, 0.0454 N solution is required.

## Reagents

- Phenolphthalein indicator solution
- Standard sodium carbonate (0.0454 N)

## Procedure

**Collection of samples:** Collection of water sample and it does not come in contact with the atmosphere. To minimise exposure to air, a portion of the sample is first siphoned into a 100 ml graduated cylinder through a flexible tube which discharges at the bottom of the cylinder. About, 50 to 75 ml of water may be allowed to overflow from the cylinder before the tube is removed. The excess sample from the cylinder is carefully removed using a pipette. The analysis for $CO_2$ should be completed within 2 or 3 hours of sample collection.

**Titration of carbon dioxide:** Add 4 drops of phenolphthalein indicator solution to 100 ml of sample. If the sample turns pink the pH is above 8.3 and free $CO_2$ is absent. If the sample remains colourless it contains free $CO_2$. Samples containing $CO_2$ must be titrated rapidly with 0.0454 N $Na_2CO_3$ solution. Stir gently with a stirring rod as $Na_2CO_3$ is added. A definite pink colour which remains for 30 seconds marks the end point.

To improve accuracy, a portion of the sample is titrated to the end point and a second 100 ml portion of the sample is taken and the volume of sodium carbonate solution titrate used in the first titration is quickly added. The titration is then continued to the end point.

## Calculation

$$CO_2\,(mg/l) = \frac{(ml\ of\ Na_2CO_3)\ x\ N\ x\ E\ x\ 1000}{Sample\ volume\ (ml)}$$

Where,

N = Normality of $Na_2CO_3$

E = Equivalent weight of $CO_2$ (22)

## Guidelines for $CO_2$ values for fish culture:

- 12-50 mg/litre- sub-lethal effects include respiratory stress and development of kidney stones (nephrocalcinosis) in some species
- 50-60mg/litre- lethal to many fish species with prolonged exposure

## Total Alkalinity

The alkalinity of normal freshwater results from bicarbonate, carbonate and hydroxide ions. Alkalinity is measured volumetrically by titration with dilute hydrochloric or sulphuric acid. Two end point indicators, methyl orange and phenolphthalein are commonly employed for alkalinity titrations. Results are expressed as mg/l equivalent of Calcium Carbonate.

## Principle

The water samples containing measurable carbonate ions turn pink to phenolphthalein indicator (pH about 8.4) and the measurement of alkalinity is done by titrating the sample in two stages. During titration to the phenolphthalein end point, all of the hydroxides and half of the carbonate are neutralised. Reactions are illustrated in equations 1 and 2.

$$OH^- + H^+ \text{----------------}> H_2O$$

$$CO_3^{2-} + H^+ \text{----------------}> HCO_3$$

Titration to the methyl orange end point (pH-4.5) neutralizes any naturally occurring bicarbonate and that bicarbonate formed by titration of carbonate (eq.3).

$$HCO_3^- + H^+ \text{--------------------}> H_2O + CO_2$$

Samples negative to phenolphthalein, but positive to methyl orange, contain only bicarbonate. Samples negative to methyl orange have a pH below 4.5 and have no alkalinity.

The entire amount of acid required to titrate a sample of water to pH 4.5 is equivalent to the total alkalinity. The titration to pH 8.3, the phenolphthalein end point, is equivalent to the phenolphthalein alkalinity. Results of the titration to the two end points may also be used to calculate hydroxide, carbonate and bicarbonate alkalinities.

Uptake and release of carbon dioxide by organisms will change the pH and the amount of carbonate, bicarbonate or hydroxide in samples. Therefore, titrations should be done as soon as possible.

## Reagents

- ☆ Dilute $H_2SO_4$ (approximately 0.02 N)
- ☆ Phenolphthalein indicator solution
- ☆ Methyl orange indicator solution
- ☆ Standard sodium carbonate (0.02 N)

## Procedure

**Standardisation of dilute sulphuric acid:** Pipette 10 ml of standard 0.02 N $Na_2CO_3$ solution into a 250 ml beaker. Add 90 ml of $CO_2$ free distilled water. Add 4 to 8 drops of methyl orange indicator solution. Select a number of drops of methyl orange which allows easy end point detection. Use this number of drops in subsequent drops. Titrate over a white surface to the methyl orange end point with dilute $H_2SO_4$ solution. At the end point, one drop of acid will change the colour of methyl orange from yellow to orange. Calculate the normality of $H_2SO_4$ by use of the equation $N_1V_1 = N_2V_2$.

**Titration of Phenolphthalein alkalinity:** Measure a 100 ml sample unto a 250 ml beaker. Add 2 drops of phenolphthalein indicator solution. If the sample turns pink, it contains phenolphthalein alkalinity. Titrate with standard $H_2SO_4$ solution until one drop of acid causes the pink colour of the sample to disappear. Save the sample for determination of methyl orange alkalinity.

**Titration of methyl orange alkalinity:** Add 4 to 8 drops of methyl orange indicator solution to the same sample used for phenolphthalein alkalinity titration. Titrate over a white surface to the methyl orange end point with standard $H_2SO_4$ solution.

## Calculation:

| | | |
|---|---|---|
| Phenolphthalein alkalinity as mg/l equivalent $CaCO_3$ | = | ml $H_2SO_4$ required for phenolphthalein end point X Normality of $H_2SO_4$ X 50 X 1000/ Volume of sample (ml) |
| Total alkalinity as mg/l equivalent $CaCO_3$ | = | Total ml $H_2SO_4$ required for methyl orange & phenolphthalein end points X normality of $H_2SO_4$ X 50 X 1000/Volume of Sample (ml) |

## Inferences

The data for phenolphthalein and total alkalinity may be used to asses hydroxide, carbonate and bicarbonate alkalinities.

- Phenolphthalein alkalinity is zero: Sample contains only bicarbonate alkalinity.
- Phenolphthalein alkalinity is less than half total alkalinity: Sample contains both carbonate and bicarbonate alkalinity.
- Phenolphthalein alkalinity x 2 = Carbonate alkalinity
- Total alkalinity – Carbonate alkalinity = Bicarbonate alkalinity
- Phenolphthalein alkalinity is half total alkalinity: Sample contains only carbonate alkalinity.
- Total alkalinity = Carbonate alkalinity
- Phenolphthalein alkalinity is more than half total alkalinity: Sample contains both hydroxide and carbonate alkalinity.

(Total alkalinity – Phenolphthalein alkalinity) X 2 = Carbonate alkalinity

Total alkalinity – Carbonate alkalinity = Hydroxide alkalinity

- Phenolphthalein alkalinity is equals total alkalinity: Sample contains only hydroxide alkalinity.

Total alkalinity = Hydroxide alkalinity

## Total Hardness

The total hardness of a water sample is defined as its content of divalent metallic cations expressed as equivalent calcium carbonate. Calcium ($Ca^{2+}$) and Magnesium ($Mg^{2+}$) are the predominant divalent metallic cations in most waters. Alkalinity and hardness values are normally similar in magnitude because calcium, magnesium, bicarbonate, and carbonate ions in water are derived in equivalent quantities from the solution of the limestone in geological deposits. However, in some water's alkalinity may exceed its hardness and vice versa. If alkalinity is high and hardness low, pH may rise to very high levels (>10.5) during rapid photosynthesis. Hardness of a water sample can be estimated by a volumetric method.

### Principle

The calcium and magnesium ions of the water sample are titrated with ethylene diamine tetra acetic acid disodium salt (EDTA) to form the stable Ca EDTA and Mg EDTA. A small quantity of Erichrome black-T added to a water sample and buffered at pH 10 would lead to a soluble wine-red complex, with some of the calcium and magnesium ions. During titration, the EDTA will first complex all of the free $Ca^{2+}$ and $Mg^{2+}$ and the solution turns blue. The calcium and magnesium would then dissociate from their complexes with Erichrome black-T to form more stable complexes with EDTA.

## Reagents

- ☆ Buffer solution
- ☆ Eriochrome black-T Indicator
- ☆ Standard calcium solution (0.01 M)
- ☆ Standard EDTA Titrant

## Procedure

A 100 ml of water sample is taken in a flask and to it added 2 ml of the buffer solution. After stirring, 8 drops of Erichrome black-T indicator are added and the solution is titrated against the EDTA solution. A colour change from wine red to pure blue is the end point. The total hardness is calculated as follows.

$$Total\ hardness\ (mg / l\ as\ CaCO_3) = \frac{(ml\ of\ EDTA)\ (M)\ (100.1)\ (1000)}{Sample\ Volume\ (ml)}$$

Chapter-6

# Water Quality Criteria/ Requirements for Aquaculture

Fishes always remains in equilibrium with their environment and potential disease-causing microbes. Alteration in this equilibrium, such as deterioration in water quality can result in fish becoming stressed and vulnerable to diseases. It may ultimately also hamper their growth. Hence, it is essential to know congenial water-quality criteria/requirements for aquaculture.

## Temperature

Fish and crustaceans are poikilotherms. The rates of all biochemical processes are temperature dependent. Within the temperature range that normally occurs in the natural habitat of a particular species, the rate of biochemical processes roughly doubles for every 100°C increase in temperature. Under favourable conditions, optimum temperature range for many coldwater and warm water fishes is 14-18$^0$C and 24-30$^0$C respectively. <12$^0$C lethal for tropical major carps but congenial for cold water fish species < 20$^0$C sub lethal for growth and survival of most of the tropical carps and > 35$^0$ C lethal to most of the fish species (Bhatnagar et al. 2004).

## Turbidity

Clay turbidity in water to 30 cm or less may prevent development of plankton blooms, 30 to 60 cm and as below 30 cm-generally adequate for good fish production and if values are > 60 cm there may be increase occurrence of low DO problems. Turbidity in the range from 30-80 cm is good for fish health; 15-40 cm good for intensive culture system and < 12 cm causes stress. Nephelometer reading of 20-30 NTU is considered as suitable for fish culture.

## Water Colour

Abundance of Phyto & Zooplankton are responsible for determination of colour of an aquatic body and green, bluish green/ brown greenish colour of water indicates good plankton population hence, good for fish health.

## Dissolved Oxygen (DO)

This is the most critical and limiting factor in aquaculture. $O_2$ enters water through photosynthesis by aquatic plants, mainly phytoplankton, and by diffusion at the air-water interface. Critical limits of DO are shown in Fig.

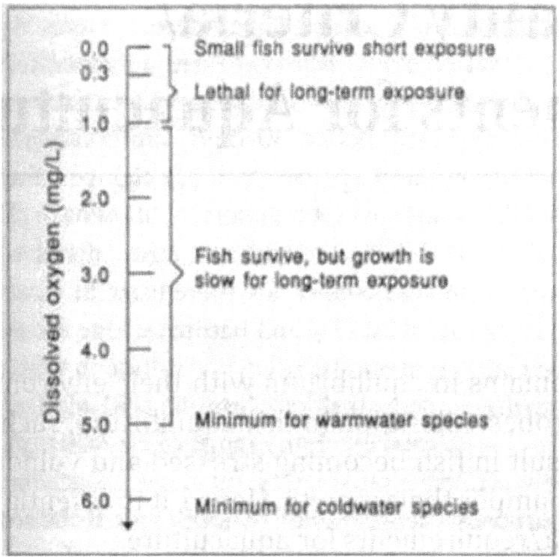


The optimum DO content of pond water should be in the range of 5mg/L to saturation level for good fish growth. Some of the guidelines for DO for fish production are:

- 5mg/litre- optimum for normal growth and reproduction in tropical waters
- 1.0-5.0 mg/litre-may have sub-lethal effects on growth, feed conversion and tolerance to disease
- 0.3-0.8 mg/litre-lethal to many species if sustained for a long period

## pH

It is a measure of hydrogen ion concentration in water and indicates how much acidic or basic the water is. $CO_2$ has an acidic reaction in water. The pH in pond rises during day because of phytoplankton and other aquatic plants remove $CO_2$ from the water during photosynthesis. It decreases at night because of respiration and production of $CO_2$ by all organisms.

Water pH affects metabolism and physiological process of fish; it also exerts considerable influence on toxicity of ammonia and hydrogen sulphide as well as solubility of nutrients and thereby water fertility. Guidelines for pH value for fish production are shown in Table:

| pH | Effect |
|---|---|
| 4 | Acid death point |
| 4-6 | Slow growth |
| 6-9 | Best for growth |
| 9-11 | Slow growth, lethal to fish over long period of time |
| 11+ | Alkaline death point |

### Free $CO_2$

$CO_2$ is present in the atmosphere in very small quantity. For this reason, although it is highly soluble in water, its concentration in most water-bodies is low. It occurs in water in 3 closely related forms: (a) free $CO_2$, (b) bicarbonate ion ($HCO_3$-) and (c) carbonate ion ($CO3^{2-}$). Each form amount present depends on the pH of water. As for example, in neutral or acidic waters, high concentration of free $CO_2$, i.e. toxic form is frequently found.

Carbon dioxide is important in pond fish culture as:

- It can be a stressor of aquatic animals
- It influences the pH of water
- It is a nutrient for plant growth and its availability may limit primary productivity of some aquatic ecosystems.

In pond fish culture, it is crucial to maintain a free carbon dioxide level of 5-8 ppm for optimal photosynthetic activity. Carbon dioxide concentrations ranging from 12-15 ppm are considered sublethal to fishes, while levels exceeding 50-60 ppm can be lethal. Elevated dissolved $CO_2$ concentrations (> 60-80 mg/L) have a necrotic effect on aquatic animals, and even higher concentrations may result in death. Exposure to lower concentrations can stress fish by interfering with respiration or inducing the formation of calcareous deposits in the kidney (Nephrocalcinosis). Hypercapnia, characterized by high environmental concentrations of dissolved $CO_2$, reduces $CO_2$ excretion at fish gills, leading to elevated plasma $CO_2$ levels and uncompensated respiratory acidosis. These conditions decrease the affinity of haemoglobin for oxygen, resulting in reduced oxygen uptake by blood at the gills, despite high environmental dissolved oxygen concentrations.

## Total Alkalinity

Alkalinity refers to buffering capacity of water. The alkalinity of normal freshwater results from bicarbonate, carbonate and hydroxide ions.

Guidelines of alkalinity for fish growth are as follows:

- 300 mg/litre- creates stress to fish
- 75-300 mg/l – ideal for fish
- < 75 mg/l- creates stress to fish

## Total Hardness

The total hardness of a water sample is defined as its content of divalent metallic captions expressed as equivalent calcium carbonate. Calcium ($Ca^{2+}$) and Magnesium ($Mg^{2+}$) are the predominant divalent metallic cations in most waters. Alkalinity and hardness values are normally similar in magnitude because calcium, magnesium, bicarbonate, and carbonate ions in water are derived in equivalent quantities from the solution of the limestone in geological deposits. However, in some water's alkalinity may exceed its hardness and vice versa. If alkalinity is high and hardness low, pH may rise to very high levels (>10.5) during rapid photosynthesis.

Waters are often categorised as follows according to degrees of hardness:

- 0-75 mg/litre- Soft
- 75-150 mg/litre-moderately hard
- 150-300 mg/litre-hard
- 300 mg/litre- very hard

## Guidelines for Hardness Value for Fish Growth

- 60 mg/litre- satisfactory for pond productivity and helps protect fish against harmful effects of pH fluctuation and metal ions
- <60 mg/litre- creates stress to fish

Alkalinity and hardness are not much affected by biological activity or aquacultural operations, and the initial concentrations in ponds are determined by their levels in the water supply; any changes are largely the result of rainfall and evaporation. Desirable level for fish culture generally fall within the range of 75-300 mg/litre.

### Ammonia

Ammonia is the by-product from protein metabolism excreted by fish and bacterial decomposition of organic matter such as waste food, faeces, dead plankton, sewage etc. The unionised form of ammonia ($NH_3$) is extremely toxic while the ionised form ($NH_4^+$) is not and both the forms together known as "Total ammonia". Un-ionised ammonia concentrations are determined by total ammonia concentration, pH and water temperature. At any given level of total ammonia, concentrations of un-ionised ammonia are greatest in surface waters on warm, sunny afternoons when pH and temperature are highest. Un-ionised ammonia concentrations decrease at night as carbon dioxide produced in respiration causes pH to decline.

The guidelines for unionized ammonia levels for fish growth are:

- ☆ 0.02-0.05 mg/litre-safe concentration for many tropical fish species
- ☆ 0.05-0.4 mg/litre- sub-lethal effects depending on the species

☆ 0.4-2.5 mg/litre-lethal to many fish species

## Nitrite

Nitrite is an intermediate product of aerobic nitrification bacterial process, produced by the autotrophic Nitrosomonas bacteria combining oxygen and ammonia. In most natural water-bodies and in well-maintained ponds, nitrite concentration is low. In water-bodies with high organic pollution and low oxygen concentration, nitrite concentration may increase.

Nitrite can be termed as an invisible killer of fish because it oxidizes haemoglobin to methaemoglobin in the blood, turning the blood and gills brown and hindering respiration also damaging nervous system, liver, spleen and kidneys of fish. This is also known as "Brown Blood Disease".

Guidelines for nitrite value for fish growth are as follows:

- ❑ 0.02-1.0 mg/litre- sub-lethal level for many fish species.
- ❑ 1.0-10 mg/litre-lethal level for many warm water fish species.
- ❑ <0.02 mg/litre is acceptable.

## Nitrate

Nitrate ($NO_3$) is harmless and is produced by the autotrophic Nitrobacter combining oxygen & nitrite. Nitrate levels are normally stabilized in the 50-100 ppm range. Nitrate is relatively nontoxic to fish and not cause any health hazard except at exceedingly high levels (>90 ppm).

## Phosphorus

Phosphorus is considered as the limiting nutrient for freshwater pond environment. Almost all the phosphorus (P) present in water is in the form of phosphate ($PO_4$) and in surface water mainly present as bound to living or dead particulate matter and in the soil is found as insoluble $Ca_3(PO_4)_2$ and adsorbed phosphates on colloids except under highly acid conditions. It is an essential plant nutrient as it is often in limited supply and stimulates plant (algae) growth and its role for increasing the aquatic productivity is well recognized.

- ❑ Phosphate level of 0.06 mg $L^{-1}$ is desirable for fish culture
- ❑ 0.05-0.07 ppm is optimum and productive
- ❑ ppm is good for plankton / shrimp production

## Hydrogen sulphide

Freshwater fish ponds should be free from $H_2S$. Hydrogen sulphide is produced by chemical reaction of organic matter that accumulates and forms a thick layer of organic deposit at the bottom. Unionised $H_2S$ is toxic to fish, but ions resulting from its dissociation are not very toxic. Guidelines for $H_2S$ value for fish growth are as follows:

- 0.01-0.5 mg/litre-lethal to fish and any detectable concentration of hydrogen sulphide in water creates stress to fish
- 0.1-0.2 mg/litre-prawns lose their equilibrium and may create sub-lethal stress;
- 3 mg/litre-prawns die instantly.

## Redox Potential (Oxidation-reduction Eh)

It is an index showing status of oxidation or reduction. It is correlated with chemical substances such as $O_2$, $CO_2$ and minerals composed of aerobic layers; $H_2S$, $CO_2$, $NH_3$, $H_2SO_4$ and others comprising anaerobic layer. Microorganisms are correlated with the status of oxidation or reduction. With the degree of Eh, it is indicative of one of the parameters that shows supporting ability of water and soil to fish and prawn biomass.

## Iron

In surface waters, iron occurs in ferrous state II (soluble compounds) or ferric state III (mostly insoluble compounds) forms. Ratio of these two forms depends on the oxygen concentration in the water, pH and other chemical properties of water. Poorly oxygenated water with low pH, where iron is mainly in the form of soluble compounds may harm fishes. As gill surface of fish tends to be alkaline, soluble ferrous iron can be oxidised to insoluble ferric compounds, which then cover gill lamellae and inhibit respiration. In cyprinid culture, it is generally accepted that concentration of iron soluble ionised forms should not exceed 0.2 mg/litre; for salmonids, the limit is 0.1 ml per litre.

## Chlorine

To control bacteria, municipal water supplies are treated with chlorine 1.0mg/litre. If municipal waters are used to culture fish, residual chlorine must be removed by aeration with chemicals such as sodium thiosulphate, or filtration through activated charcoal. Chlorine level as low as 0.02 mg/litre can stress fish.

## Salinity

Salinity refers to the total concentration of all ions in water. The major ions contributing to salinity are calcium, magnesium, sodium, potassium, bicarbonate, chloride and sulphate. Ground waters of low salinity (<0.1ppt) are found in inland aquifers composed of sparingly soluble silicate minerals. Limestone aquifers typically yield waters of some higher salinity (0.1-1 ppt), with calcium, magnesium, and bicarbonate being the major ions. High salinities (>15ppt) are typical of groundwaters in contact with halite (NaCl), gypsum ($CaSO_4$), or other soluble minerals. Coastal plain aquifers (and some inland aquifers) derived from marine sediments or subject to seawater intrusion also may have relatively high salinities (from 5 to 35 ppt), with sodium, chloride, and bicarbonate being the major ions.

Fishes are sensitive to the salt concentration of their waters and have evolved a system that maintains a constant salt ionic balance in its bloodstream through the movement of salts and water across their gill membranes.

**Water quality parameters and acceptable ranges for pond management**

| S. No | Parameter | Acceptable range | Desirable range | Stress |
|---|---|---|---|---|
| | Temperature ($^0$C) | 15-35 | 20-30 | <12, >35 |
| | Turbidity (cm) | 12-30 | 30-80 | <12, >80 |
| | Water colour | Pale to light green | Light green to light brown | Clear water, Dark green & Brown |
| | DO (ppm) | 3-5 | 5 | <5, >8 |
| | BOD (ppm) | 3-6 | 1-2 | >10 |
| | Free $CO_2$ (ppm) | 0-10 | <5, 5-8 | >12 |
| | pH | 7-9.5 | 6.5-9 | <4,>11 |
| | Total alkalinity (ppm) | 50-200 | 75-300 | <20, >300 |
| | Total hardness (ppm) | >20 | 75-300 | <20, >300 |
| | Ca (ppm) | 4-160 | 25-100 | <10, >250 |
| | $NH_3$ (ppm) | 0-0.05 | 0 - <0.025 | >0.3 |
| | $NO_2$ (ppm) | 0.02-0.2 | <0.02 | >0.2 |
| | $NO_3$ (ppm) | 0-100 | 0.1-4.5 | >100, <0.01 |
| | $PO_4$-P (ppm) | 0.03-2 | 0.01-3 | >3 |
| | $H_2S$ (ppm) | 0-0.02 | 0.002 | Any detectable level |

Chapter-7

# Soil Concepts and Characteristics of Soils

## Origin and Nature of Soils

- The term SOIL was derived from the latin word "SOLUM" means FLOOR. For a layman soil is dirt or debris. For an Agriculturist soil is a habitat for plant growth (to grow crops). For a mining engineer soil is debris covering the rocks. For a civil engineer soil is a material on which road bed or house bed is formed. For a home owner soil is a mellow or loamy or hard material.
- Soil is a three-dimensional body having length, breadth and depth. They form a continuation over the land surface and differ in properties from place to place.
- Its upper boundary is air or water and lower boundary is the rock lithosphere.
- Ruffin and Simonson (1968): Soil is a mixture of earth's uppermost mantle of weathered rock and organic matter.
- Buckman and Brady (1969): Soil is a dynamic natural body on the surface of the earth in which plants grow, composed of mineral and organic materials and living forms.

## Definition of Soil

- Generally, soil refers to the loose surface of the earth as identified from the original rocks and minerals from which it is derived through weathering process.

- **Dokuchaiev (1900):** Russian scientist - Father of soil science - Soil is a natural body composed of mineral and organic constituents, having a definite genesis and a distinct nature of its own.

## Soil Formation

### Weathering

- A process of disintegration and decomposition of rocks and minerals which are brought about by physical agents and chemical processes, leading to the formation of Regolith (unconsolidated residues of the weathering rock on the earth's surface or above the solid rocks) (OR)
- The process by which the earth's crust or lithosphere is broken down by the activities of the atmosphere, with the aid of the hydrosphere and biosphere (OR)
- The process of transformation of solid rocks into parent material or Regolith

**Parent material:** It is the regolith or at least it's upper portion. May be defined as the unconsolidated and more or less chemically weathered mineral material, from which soil are developed.

### Weathering can be of 3 types

- **Physical weathering** - breakdown of rocks from the result of a mechanical action like temperature changes, abrasion (when rocks collide with each other) or frost can all cause rocks to break down.
- **Chemical weathering** - breakdown of rocks through a change in their chemical makeup. This can happen when the minerals within rocks react with water, air or other chemicals.
- **Biological weathering** - the breakdown of rocks by living things. Burrowing animals help water and air get into rock, and plant roots can grow into cracks in the rock, making it split.

### Factors influencing soil formation

- Climatic condition - particularly temperature and precipitation
- Living organism - especially native vegetation
- Nature of parent material – texture & structure of soil
- Chemical & mineralogical composition of soil
- Topography of the area
- Time of soil formation

## Physical Properties of Soil

- Soil is a complex mixture of minerals, organic matter, liquids, gases, and microorganisms that all work together to support life.

- Soil is composed of four main components: minerals, organic matter, water, and air.
- These components are arranged in distinct layers, or horizons, which make up the soil profile
- The top layer of soil is called the topsoil, which is rich in organic matter and supports the majority of plant life.
- Below the topsoil is the subsoil, which contains more minerals and less organic matter.
- The deepest layer of soil is called the parent material, which is composed of weathered rock fragments and minerals. The soil properties includes;

## Soil Colour

The colour of soil, whether brown, yellow, or red, is determined by the presence of oxidized or ferric iron compounds. Darker soil hues typically indicate greater organic content. With increased organic matter, soil tends to absorb more heat, resulting in higher temperatures.

## Soil Structure

Soil structure refers to the organization of soil separates into units known as soil aggregates, which consist of solids and pore spaces. These aggregates are separated by planes of weakness and are primarily composed of clay particles, although silt and fine sand particles may also be present. Depending on their size, aggregates behave like larger silt or sand particles.

The arrangement of soil aggregates into various forms contributes to the structure of soil. Natural processes that facilitate aggregate formation include:

- Wetting and drying
- Freezing and thawing
- Microbial activity aiding in organic matter decay
- Root and soil animal activity
- Adsorbed cations

These processes, such as wetting/drying cycles and freezing/thawing action, as well as root or animal activity, promote the movement of particles, leading to aggregate formation. Decomposing plant residues and microbial byproducts coat soil particles, binding them into aggregates. Additionally, adsorbed cations contribute to aggregate formation by bonding with multiple particles. Aggregates are characterized by their shape, size, and stability, with aggregate types being commonly discussed when addressing soil structure.

| Type | Description |
|---|---|
| Granular | Rounded surfaces |
| Crumb | Rounded surfaces but larger than granular |
| Subangular blocky | Cube-like with flattened surfaces and rounded corners |
| Blocky | Cube-like with flattened surfaces and sharp corners |

| Type | Description |
|---|---|
| Prismatic | Rectangular with a long vertical dimension and flattened top |
| Columnar | Rectangular with a long vertical dimension and rounded top |
| Platy | Rectangular with a long horizontal dimension |
| Single grain | No aggregation of course particles when dry |
| Structureless | No aggregation of fine particles when dry |

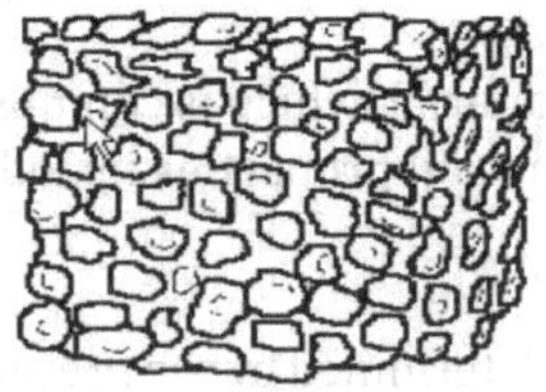

**Granular**: Resembles cookie crumbs and is usually less than 0.5 cm in diameter. Commonly found in surface horizons where roots have been growing.

**Blocky**: Irregular blocks that are usually 1.5 - 5.0 cm in diameter.

**Prismatic**: Vertical columns of soil that might be a number of cm long. Usually found in lower horizons.

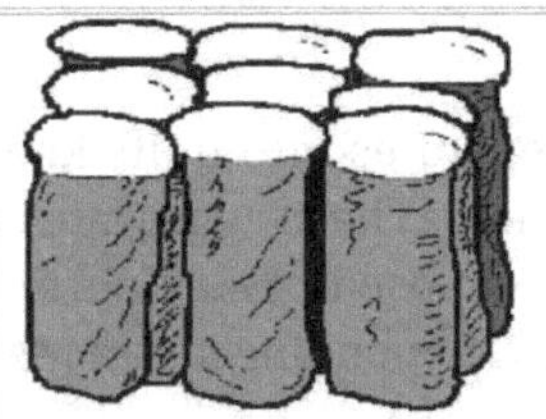

**Columnar**: Vertical columns of soil that have a salt "cap" at the top. Found in soils of arid climates.

**Platy**: Thin, flat plates of soil that lie horizontally. Usually found in compacted soil.

Soil Science Society of America

**Single Grained**: Soil is broken into individual particles that do not stick together. Always accompanies a loose consistence. Commonly found in sandy soils.

## Soil Texture

- Soil texture refers to the proportion of sand, silt, and clay-sized particles constituting the mineral fraction of the soil, determining classifications like loam, sandy loam, or clay.
- Sand and silt are considered less significant in soil composition as they do not significantly contribute to the soil's capacity for retaining water and nutrients.
- Clay, on the other hand, plays an active role in soil texture due to its minute size and expansive surface area per unit mass, facilitating water and ion retention.
- Soil texture serves as an indicator of the soil's water-holding capacity, the rate of water infiltration, as well as its workability and fertility.

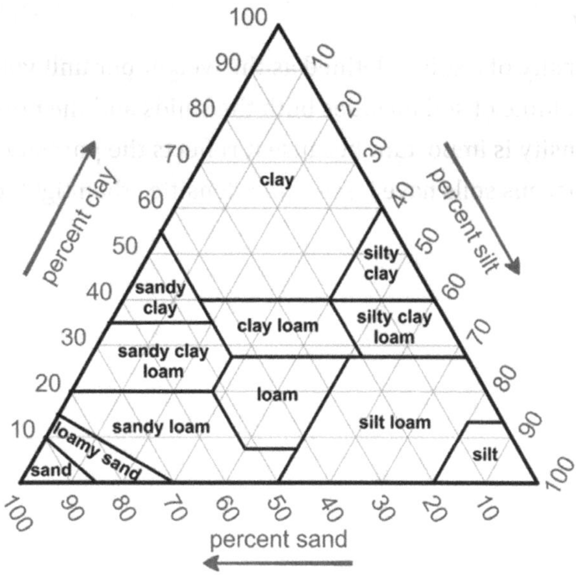


## Water Absorption

- Because of its porosity, soil can absorb water.
- The capacity of different soil types to hold water varies.
- Sand absorbs significantly less water than clay. Sandy soil has a lower water-holding capacity than clay and loamy soil.
- Clay soil is more water-retaining than sandy soil.

## Porosity

- Porosity refers to the space between particles in the soil.
- Pores aid in the retention of air, water, and microorganisms.
- Porous soil is soil that has a large pore.
- Because sandy soil has half the porosity of clay soil, clay soil has a higher water or air-holding capacity than sandy soil.
- More porosity means more capacity to hold water or air.

## Soil Permeability

- Permeability refers to the rate at which moisture or air passes through the soil.
- Water flows easily through sandy soil but not clay soil; thus, sandy soil is more permeable than clay soil.
- The soil with the most pores is the least permeable

## Bulk Density

- Bulk density of a soil is defined as the weight per unit volume of soil.
- A unit volume of soil includes both the solids and the pore space.
- Bulk density is important because it reflects the porosity of a soil.
- Loose, porous soils have lesser bulk densities than tight, compacted soil.

Chapter-8

# Soil Types and their Distribution

## Soil Types

Soil can be classified into four types namely; Sand, Silt, Clay and Loam.

### Sandy Soil

The first type of soil is sand. They consist of small particles of weathered rocks. They are one of the poorest types of soil for growing plants as they are low in nutrient and have poor water holding capacity. They are usually formed as a result of the breakdown and fragmentation of rocks like granite, limestone and quartz.

### Silt Soil

They are comparatively smaller particles in comparison with sandy soil and is made up of rocks and other mineral particles. They are smaller than sand but larger than clay.

### Clay Soil

Clay is the smallest particle amongst the types of soil. They are tightly packed against each other with little or no air space. They have a strong water holding capacity. Clay is the densest type of soil which does not drain well or provide space for plant roots to flourish.

## Loamy Soil

It is a combination of sand, silt and clay such that the beneficial properties from each are included. For instance, their ability to retain moisture, nutrients. This soil is also referred to as the agricultural soil.

The United States Department of Agriculture System (USDAS) and International Soil Science Society System (ISSSS) categorized soil based on size as:

| SL. NO | SOIL TYPE | SIZES (mm) | |
|---|---|---|---|
| | | USDAS | ISSSS |
| 1. | Very coarse sand | 2.00-1.00 | - |
| 2. | Coarse sand | 1.00-0.50 | 2.00-0.20 |
| 3. | Medium sand | 0.50-0.25 | - |
| 4. | Fine sand | 0.25-0.10 | 0.20-0.02 |
| 5. | Very fine sand | 0.10-0.05 | - |
| 6. | Silt | 0.05-0.002 | 0.02-0.002 |
| 7. | Clay | <0.002 | <0.002 |

## Soil Profile

The vertical cross-section of the soil, made of layers running parallel to the surface is called soil profile. Each layer of the profile is known as soil horizon.

The different layers of soil are:

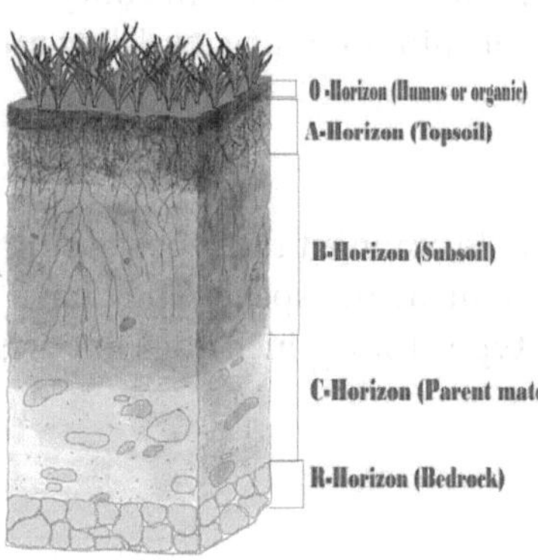

## O-Horizon

- The O horizon is the upper layer of the topsoil which is mainly composed of organic materials such as dried leaves, grasses, dead leaves, small rocks, twigs, surface organisms, fallen trees, and other decomposed organic matter.
- This horizon of soil is often black brown or dark brown in colour and this is mainly because of the presence of organic content.

## Topsoil or A horizon

- It is the darkest layer of the soil because it has the highest proportion of organic material and is known as the humus layer.
- This layer consists of both organic matter and other decomposed materials.
- The topsoil is soft and porous to hold enough air and water.
- In this layer, the seed germination takes place and new roots are produced which grows into a new plant.
- This layer consists of microorganisms such as earthworms, fungi, bacteria, etc.

## Subsoil or B horizon

- It is the subsurface horizon, present just below the topsoil and above the bedrock.
- It is comparatively harder and compact than topsoil.
- It contains less humus, soluble minerals, and organic matter.
- It is a site of deposition of certain minerals and metal salts such as iron oxide.
- This layer holds enough water than the topsoil and is lighter brown due to the presence of clay soil.
- The A and B horizons together are referred to as the Solum or True Soil

## C-Horizon or Saprolite

- This layer is devoid of any organic matter and is made up of broken bedrock.
- This layer is also known as saprolite.
- The geological material present in this zone is cemented.

## R-Horizon

- It is a compacted and cemented layer.
- Different types of rocks such as granite, basalt and limestone are found here.

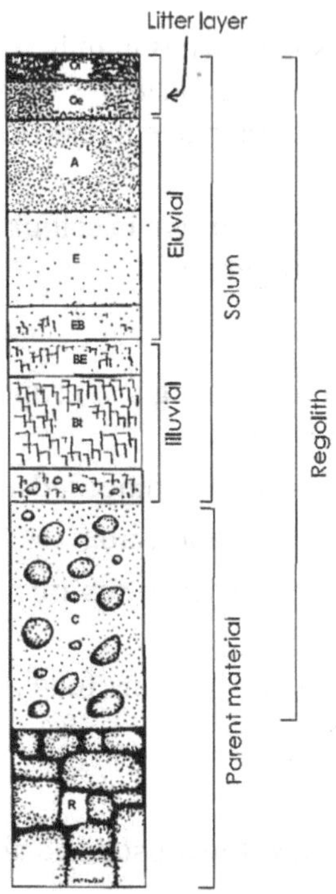


**O:** litter, forest floor; Oi common, ½-1" thick;
Oa,Oe usually very thin or absent

**A:** topmost mineral soil; 2-12" typical;
use "p" if abrupt boundary (tillage mark)

**E:** often absent due to erosion; may be light-colored,
usually same texture as A; not blocky

Transition: usually only one– BE or EB; B often first,
since more distinct

**B:** usually clayey, blocky, Fe coloration; 6-36" thick;
may be subdivided (B1,B2); use t, w, or h–
only one type per profile

Transition: usually BC (B more distinct)

**C:** weathered parent material; lower clay,
massive/rock structure; if clear rock-like structure
and appreciable rock fragments, use r ("Cr")

**R:** hard rock (cannot be dug); describe type if possible

## Types of Soil in India

India has several different soil types, including alluvial, black, red, laterite, desert, and mountain soils.

### Alluvial Soil

Alluvial soil is the most common soil type in India, and is found in the Indo-Gangetic plains, the Brahmaputra valley, and the coastal plains. This soil is formed by the deposition of silt and clay by rivers, and is rich in nutrients, making it ideal for agriculture.

### Black Soil

Black soil, also known as regur soil, is found in the Deccan plateau region of India. This soil is rich in iron, magnesium, and other nutrients, and is known for its moisture retention properties, making it ideal for crops like cotton and tobacco.

### Red Soil

Red soil is found in the southern and eastern parts of India, including Tamil Nadu, Andhra Pradesh, and Odisha. This soil is rich in iron and aluminum, and is low in fertility, making it less suitable for agriculture.

## Laterite Soil

Laterite soil is found in the hilly areas of western and eastern ghats, parts of Kerala, Karnataka, and Maharashtra. This soil is characterized by its red colour, high iron content, and poor fertility, making it unsuitable for agriculture.

## Desert Soil

Desert soil is found in the Thar desert region of India, and is characterized by its sandy texture and low fertility. This soil is used for grazing livestock and is unsuitable for agriculture.

## Mountain Soil

Mountain soil is found in the Himalayan region of India, and is characterized by its rocky texture and low fertility. This soil is used for horticulture and plantation crops.

Chapter-9

# Soil Chemistry

## Soil Colloids

Soil colloids are tiny particles found in soil that play a crucial role in soil fertility and nutrient availability. They are composed of both organic and inorganic matter and have a high surface area relative to their volume. The main types of soil colloids are clay particles, organic matter, and iron and aluminum oxides.

## Types of Soil Colloids

- Clay Colloids: Clay colloids are the smallest soil particles, typically with a diameter less than 0.002 millimetres. They are primarily composed of silicate minerals such as kaolinite, montmorillonite, and illite. Clay colloids have a high surface area and a net negative charge, which enables them to attract and retain positively charged ions (cations) such as potassium ($K^+$), calcium ($Ca^{2+}$), and magnesium ($Mg^{2+}$). This property is essential for soil fertility and nutrient retention.
- Organic Colloids: Organic colloids are derived from decomposed plant and animal matter in the soil. They include humus, microbial residues, and other organic substances. These colloids play a crucial role in enhancing soil fertility and structure by enhancing water retention, increasing nutrient availability, and promoting soil aggregation. Similar to clay colloids, organic colloids bear a negative charge, enabling them to attract and retain cations, thus serving as a nutrient reservoir for plants.
- Iron and Aluminum oxide colloids: These colloids are formed from the weathering of minerals containing iron (Fe) and aluminum (Al).

They include minerals such as iron oxides (e.g., hematite, goethite) and aluminum oxides (e.g., gibbsite). Iron and aluminum oxide colloids have a high surface area and a strong negative charge, which enables them to adsorb and retain nutrients, organic matter, and other substances in the soil. They also contribute to soil coloration and influence soil properties such as drainage and pH.

## Properties of Soil Colloids

- **High Surface Area:** Soil colloids have an enormous surface area relative to their volume due to their small particle size. This high surface area allows soil colloids to adsorb and retain large quantities of water, nutrients, organic matter, and other substances. This property is crucial for soil fertility and plant nutrition.
- **Cation Exchange Capacity (CEC):** Cation exchange capacity is a measure of a soil's ability to adsorb and retain positively charged ions (cations) such as potassium ($K^+$), calcium ($Ca^{2+}$), magnesium ($Mg^{2+}$), and ammonium ($NH_4^+$). Soil colloids, particularly clay and organic colloids, contribute significantly to CEC due to their negative charge, which attracts and holds onto cations. CEC influences nutrient availability to plants and soil fertility.
- **Anion Exchange Capacity:** While less common than cation exchange, some soil colloids also have the ability to adsorb and retain negatively charged ions (anions) such as nitrate ($NO_3^-$) and phosphate ($PO_4^{3-}$). This property, known as anion exchange capacity, can affect the availability and leaching of anions in the soil.
- **Negative Charge:** Soil colloids typically carry a negative charge on their surfaces due to the presence of hydroxyl groups ($OH^-$) and other functional groups. This negative charge arises from the dissociation of acidic functional groups and the isomorphic substitution of cations within the crystal lattice of clay minerals. The negative charge enables soil colloids to attract and retain cations, organic matter, and nutrients through electrostatic interactions.
- **pH Buffering Capacity:** Soil colloids, especially organic colloids, contribute to the soil's pH buffering capacity. They can absorb and release hydrogen ions ($H^+$) in response to changes in soil pH, helping to stabilize soil pH and prevent rapid fluctuations. This property is essential for maintaining optimal soil pH for plant growth and nutrient availability.
- **Soil Structure and Aggregation:** Soil colloids play a crucial role in soil structure and aggregation by binding soil particles together into aggregates. This enhances soil porosity, water infiltration, aeration, and root penetration. Proper soil aggregation facilitated by soil colloids improves soil tilth and fertility.
- **Cation exchange:** Cation exchange is a fundamental process in soil chemistry that plays a crucial role in nutrient availability and soil

fertility. It involves the reversible exchange of positively charged ions, or cations, between the soil particles and the soil solution. Here's a detailed explanation of cation exchange in soil chemistry.

## Mechanism of Cation Exchange

- **Soil Colloids**: Cation exchange primarily occurs on the surfaces of soil colloids, which include clay minerals, organic matter, and iron/aluminum oxides. These colloids have charged surfaces due to isomorphic substitution, pH-dependent charges, and functional groups.
- **Charged Sites**: The charged surfaces of soil colloids attract and hold onto cations through electrostatic forces. The predominant cations involved in cation exchange include calcium ($Ca^{2+}$), magnesium ($Mg^{2+}$), potassium ($K^+$), and ammonium ($NH_4^+$).
- **Dynamic Equilibrium**: Cation exchange is a dynamic process that reaches equilibrium between the cations adsorbed onto soil colloids and those in the soil solution. When a cation is adsorbed onto a soil colloid, it can be exchanged with other cations in the soil solution.

### Factors Influencing Cation Exchange

- **Cation Exchange Capacity (CEC)**: CEC is a measure of the soil's ability to adsorb and retain cations. It depends on the type and amount of soil colloids present in the soil. Soils with higher CEC can hold more cations, which is essential for nutrient retention and availability.
- **Selectivity**: Different soil colloids have varying affinities for cations based on their charge and size. Generally, cations with higher valence (e.g., $Ca^{2+}$, $Mg^{2+}$) and smaller size have stronger adsorption onto soil colloids compared to monovalent cations (e.g., $K^+$, $Na^+$).
- **Soil pH**: Soil pH influences cation exchange by affecting the charge on soil colloids. In acidic soils, soil colloids tend to have a higher net positive charge, leading to stronger adsorption of cations. Conversely, in alkaline soils, soil colloids may have a lower net positive charge, affecting cation exchange.
- **Organic Matter Content**: Organic matter contributes to cation exchange capacity and can enhance nutrient retention and availability. Organic matter contains functional groups that can adsorb cations and hold onto them, contributing to soil fertility.

## Organic Carbon

Soil organic carbon (SOC) refers to the carbon stored within the organic matter present in soil. It is a fundamental component of soil organic matter (SOM), which includes various forms of organic materials derived from the decomposition of plant residues, animal remains, and microbial biomass.

## Formation

- **Input of organic materials:** SOC accumulates through the deposition of organic materials onto the soil surface or through root exudation. These organic inputs include plant litter, crop residues, animal manure, and microbial biomass.
- **Decomposition:** Once organic materials are deposited in the soil, they undergo decomposition by soil microorganisms such as bacteria, fungi, and other decomposers. During decomposition, organic carbon is broken down into simpler compounds, releasing energy and nutrients that support microbial growth and activity.
- **Stabilization:** Some fraction of the decomposed organic carbon becomes stabilized in the soil as humus or recalcitrant organic matter. This stabilized carbon may be physically protected within soil aggregates or chemically bound to soil minerals, making it resistant to further decomposition.

## Importance

- **Nutrient Cycling:** SOC serves as a source of energy and nutrients for soil microorganisms involved in decomposition and nutrient cycling processes. Microorganisms mineralize organic carbon, releasing nutrients such as nitrogen, phosphorus, and sulphur, which are essential for plant growth and productivity.
- **Soil Structure:** SOC plays a crucial role in soil aggregation and structure formation. It acts as a binding agent, promoting the formation of stable soil aggregates. Soil aggregates improve soil porosity, water infiltration, and root penetration, enhancing soil aeration, drainage, and nutrient availability.
- **Water Retention:** SOC enhances the soil's water-holding capacity by increasing water retention and reducing water runoff. Organic matter acts like a sponge, absorbing and holding onto water molecules, which are then available for plant uptake during dry periods or drought conditions.
- **Carbon Sequestration:** SOC contributes to carbon sequestration, the process by which atmospheric carbon dioxide ($CO_2$) is captured and stored in soil organic matter. Increasing SOC levels helps mitigate climate change by removing $CO_2$ from the atmosphere and storing it in the soil, thereby reducing greenhouse gas concentrations.

## C-N ratio

It is defined as the ratio of the weight of organic carbon to the weight of total nitrogen in a soil or organic matter. It is the relationship between organic matter and nitrogen content of soils or plants.

## Importance of C:N ratio

Importance of the C: N (Carbon to Nitrogen) ratio lies in its significant role in various biological processes, particularly in composting, soil fertility, and nutrient cycling.

- **Microbial Activity**: The C:N ratio influences microbial activity in decomposition processes. Microorganisms require carbon for energy and nitrogen for protein synthesis. A balanced C:N ratio in organic matter promotes efficient microbial decomposition, leading to the breakdown of organic materials into stable humus and the release of nutrients.
- **Composting**: In composting, it's essential to maintain an optimal C:N ratio for effective decomposition of organic materials. A balanced C:N ratio, usually around 25-30:1, promotes efficient decomposition and the generation of high-quality compost. When the C:N ratio is too high (excessive carbon), decomposition slows down, resulting in an accumulation of organic materials. Conversely, if the ratio is too low (excessive nitrogen), nitrogen may be lost through volatilization as ammonia, causing nutrient depletion and unpleasant odors.
- **Soil Fertility**: The C:N ratio affects soil fertility by influencing the availability of nutrients to plants. Organic materials with a balanced C:N ratio decompose into humus, which improves soil structure, water retention, and nutrient availability. High-quality humus formed from organic matter with an optimal C:N ratio acts as a reservoir of nutrients, releasing them gradually as they are needed by plants.
- **Nitrogen immobilization**: When the C:N ratio of organic matter is high (excessive carbon), microorganisms may immobilize nitrogen from the soil as they decompose organic materials. This temporary tie-up of nitrogen can lead to nitrogen deficiency in plants until decomposition is complete and nitrogen becomes available again.
- **Plant Growth and Productivity**: Adequate nitrogen availability, influenced by the C:N ratio, is essential for plant growth and productivity. A balanced C:N ratio in soil organic matter ensures a steady supply of nitrogen to plants, supporting healthy growth and higher yields.
- **Environmental Impact**: Maintaining an appropriate C:N ratio in agricultural practices helps reduce environmental pollution. Improper management of organic materials with imbalanced C:N ratios can lead to nutrient runoff, leaching, and eutrophication of water bodies.

## Soil Fertility

Fertile soil results in high yield and better quality of plants. Fertile soil is rich in fundamental elements and minerals, has good aeration, water holding capacity, and good texture.

## Factors Affecting Soil Fertility

The following factors affect the soil fertility

- **Soil pH:** Soil pH, which measures the acidity or alkalinity of soil, significantly influences nutrient availability. Different plants thrive in soils with specific pH ranges, and soil pH affects the solubility of essential nutrients. Most plants prefer slightly acidic to neutral soils, with a pH range of 6.0 to 7.5. Outside this range, nutrient availability can be limited, leading to deficiencies or toxicities.
- **Soil Texture and Structure:** Soil texture refers to the relative proportions of sand, silt, and clay particles in the soil. Soil structure refers to the arrangement of these particles into aggregates or clumps. Both texture and structure affect water retention, aeration, drainage, and root penetration. Sandy soils have larger particles and drain quickly but may lack nutrient retention, while clay soils have smaller particles and hold water and nutrients more effectively but may be poorly aerated.
- **Organic Matter Content:** Organic matter in soil, derived from plant and animal residues, improves soil fertility in several ways. It provides a source of nutrients as it decomposes, enhances soil structure, promotes microbial activity, increases water retention, and reduces erosion. Soils with higher organic matter content generally have better fertility and productivity.
- **Nutrient Content:** Soil fertility depends on the availability of essential nutrients such as nitrogen (N), phosphorus (P), potassium (K), calcium (Ca), magnesium (Mg), sulfur (S), and micronutrients like zinc (Zn), iron (Fe), manganese (Mn), copper (Cu), and boron (B). Nutrient availability is influenced by factors such as soil pH, organic matter content, texture, and microbial activity.
- **Biological Activity:** Soil organisms, including bacteria, fungi, earthworms, nematodes, and other microorganisms, play vital roles in nutrient cycling, organic matter decomposition, nitrogen fixation, and soil aggregation. These organisms break down organic matter, release nutrients, enhance soil structure, and suppress pathogens, contributing to soil fertility and ecosystem health.
- **Climate and Weathering:** Climate influences soil formation processes, including weathering of parent material, organic matter decomposition rates, and nutrient cycling. Temperature, precipitation, and humidity affect microbial activity, nutrient availability, and soil moisture levels, ultimately impacting soil fertility.
- **Land use and management practices:** Agricultural practices such as tillage, crop rotation, cover cropping, fertilization, irrigation, and pesticide use can affect soil fertility. Sustainable soil management practices aim to preserve soil fertility, minimize erosion, improve organic matter content, and optimize nutrient cycling, contributing to long-term agricultural productivity and environmental sustainability.

## Soil Reaction

Soil pH or soil reaction is a measure of the acidity or alkalinity of soil, indicating the concentration of hydrogen ions ($H^+$) in the soil solution. The pH scale ranges from 0 to 14, with 7 being neutral. Values below 7 indicate acidic conditions, while values above 7 indicate alkaline conditions. Soil pH plays a critical role in nutrient availability, microbial activity, and overall soil health.

**Acidity:** Acidic soils have pH values below 7, which can be attributed to various factors such as high concentrations of hydrogen ions, aluminum toxicity, or the presence of organic acids. Soil acidity can negatively affect plant growth by limiting the availability of essential nutrients such as phosphorus, calcium, and magnesium. Additionally, aluminum toxicity in acidic soils can inhibit root growth and nutrient uptake, leading to stunted plant growth and decreased crop yields.

**Alkalinity:** Alkaline soils have pH values above 7, often due to the presence of basic ions such as calcium ($Ca^{2+}$), magnesium ($Mg^{2+}$), or carbonate ($CO_3{}^{2-}$). While alkaline soils may offer some benefits such as reduced availability of toxic elements like aluminum, they can also pose challenges for plant growth by limiting the availability of certain nutrients like iron, manganese, and phosphorus. Additionally, alkaline soils may lead to nutrient imbalances and deficiencies in plants, affecting overall productivity.

**Conductivity:** Soil conductivity refers to its ability to conduct electrical current, which is influenced by factors such as soil texture, moisture content, and the concentration of ions in the soil solution. Electrical conductivity is typically measured using an electrical conductivity (EC) meter and is expressed in units such as deciSiemens per meter (dS/m) or milliSiemens per centimeter (mS/cm). High soil conductivity levels may indicate excessive salts in the soil, which can negatively impact plant growth by causing osmotic stress, reducing water uptake, and disrupting nutrient balance. Soil salinity can be managed through practices such as proper irrigation management, drainage improvement, and soil amendments.

**Redox Potential:** Redox potential, also known as oxidation-reduction potential, measures the soil's ability to transfer electrons between substances, indicating the soil's degree of oxidation or reduction. Redox potential is influenced by factors such as soil moisture, aeration, organic matter decomposition, and microbial activity. In aerobic (oxygen-rich) conditions, redox potential values are typically positive, promoting nutrient cycling, organic matter decomposition, and plant growth. However, in anaerobic (low-oxygen) environments, redox potential values may become negative, leading to processes such as denitrification, sulphate reduction, and iron and manganese reduction. These anaerobic processes can affect nutrient availability, soil structure, and the mobility of certain contaminants. Monitoring redox potential is essential for understanding soil processes, managing soil fertility, and mitigating environmental impacts such as groundwater contamination and greenhouse gas emissions.

Chapter-10

# Submerged Soil

Submerged soil are those soils that have been saturated with water for a prolonged period of time annually which gives the soil the distinctive grey horizons as a result of oxidation-reduction processes:

- Partially oxidised A horizon high in organic matter.
- Mottled zone in which oxidation and reduction alternate
- Permanently reduced zone which is bluish green in colour.

## Wet Land Soil

Wetland soils differ from terrestrial soils in that they are anaerobic. The absence of oxygen produces characteristic differences especially in colour and texture. Wetland soils are composed mostly of organic matter as the dead organic matter does not decompose well under water.

## Peat Soils

The organic materials that accumulate in swamps and marshes are called peat. Peat is composed primarily of the partially decomposed remains of wetland plants, along with lesser amounts of mineral material. It is usually olive green to black in colour in its natural state and often exhibits an odorous smell. Peat, sometimes known as turf, is an accumulation of partially decayed vegetation or organic matter. Peat soil can be categorized into three classes based on its decomposition state as sapric, hemic, and fibric.

**Sapric** materials are the most decomposed and are defined those having less than 1/6 of the total organic content.

**Fibric** materials are those having ¾ or more identifiable plant materials

**Hemic** are those that does not meet the requirement of both either sapric or fibric material.

Types of peat soils. There are 3 types of peat soils namely

- ☆ Sedimentary peat
- ☆ Fibrous peat
- ☆ Woody peat

**Sedimentary peat:** Sedimentary peats are those that usually accumulates in deeper water. They are the fine textured matrix of olive green to grey colour with little to no identifiable plant remains. As their name suggest, sedimentary peat is deposited in lakes or deeper ponds. The organic content varies but it often exceeds 50% by weight. Sedimentary peat is often soft, spongy and somewhat gelatinous in its natural wet state. Moisture capacity is as high as 4-5 times of dry weight. It has low shear strength and high consolidation potential.

**Fibrous peat:** Fibrous peat is the source of sedimentary peat and the organic portions of sand, silt, and clay. It is dark brown in colour and is composed of the remains of aquatic, semi aquatic and shore line emergent plant growth. It is deposited as shoreline deposits in ponds, lakes, and wet basins. It has very low shear strength and high consolidation potential.

**Woody peat:** Woody peat is less common than fibrous peat but is found in conjunction with that material. It is a swampy deposit accumulating at the surface above the fibrous peat. It is brown or black in colour. It is formed as emergent woody vegetations become established either on concentration or accumulation of fibrous peat or directly in shallow wet basins.

**Physical Properties of Peat Soils**

| S. No | Properties | | Description |
|---|---|---|---|
| 1. | Colour | : | Dark Brown |
| 2. | Bulk density | : | Lower than Mineral soil |
| 3. | Water holding capacity | : | High 2-3 times |
| 4. | Structure | : | Highly adsorptive, low plasticity & highly porous and Easy for cultivation |
| 5. | Buffering capacity | : | Residential to pH change |
| 6. | Lime loss | : | High as cation is very active |
| 7. | Strength of acid | : | Lower pH |
| 8. | Colloidal nature | : | Surface area increases due to soil |

## Uses of Peat Soils

- ❑ As a manure – used in nurseries, green house and lawns
- ❑ As a bedding material-used in stables litters, poultry houses,
- ❑ As a packing material-as it is a good insulation capacity.

- As a fuel-as rich in carbon coal like Dutch, Germen peat.
- Industrial purpose
- Field soils- United States for soil amendments.

## Bog Lime or Marl

- Bog lime or Marl is a grey to white deposit of fresh calcium carbonate occurring as a bottom deposit in lakes or marshes.
- It often contains some soft shell and some plant remains and generally exhibits an odorous smell.
- Marl is formed as a precipitate from the water having high calcium carbonate content.
- Marl is formed due to the deposition and decomposition of molluscs shells, aquatic plants such as mosses, algae (Chara).
- Marl may be dugout, dried and pulverized and used as agricultural lime.
- It is an alkaline peat soil and it is not desirable for aquaculture activities.

## Fluxes Between Muck and Peat Soil

Muck soil is easily identified by their black colour as a result of the high concentration of tannins. Muck soils are made up of 20-80% organic matter. Soil fertility is tied directly to the decomposition of plant residues, dead microbes, and detritus. Because of the presence of greater composition of organic matter, muck soil is more fertile than mineral soils. Muck soil also consist higher percentage of mineral matter than peat because of the decomposition of organic material. The oxidized material is called muck to distinguish it from the unoxidized peat.

## Flux Between Mud and Water

Fluxes between mud and water refer to the dynamic exchange of substances, including nutrients, gases, and pollutants, between submerged soil (mud) and the adjacent water body. This process is influenced by various factors such as soil composition, water flow dynamics, and biological activity. Here's a more elaborate explanation:

### Nutrient Fluxes

Submerged soils act as important nutrient reservoirs, storing and releasing essential elements such as nitrogen, phosphorus, and potassium. Nutrient fluxes occur as these elements are exchanged between the soil and water column through processes like diffusion, advection (water movement), and plant uptake. Nutrient fluxes play a critical role in regulating water quality, supporting aquatic plant growth, and sustaining aquatic ecosystems.

### Gas Fluxes

Gases, including oxygen ($O_2$), carbon dioxide ($CO_2$), methane ($CH_4$), and hydrogen sulphide ($H_2S$), undergo fluxes between submerged soils and water.

Oxygen is consumed through microbial respiration and oxidation reactions in the soil, leading to hypoxic (low oxygen) or anoxic (oxygen-free) conditions. Carbon dioxide is produced during organic matter decomposition, while methane and hydrogen sulphide are generated under anaerobic (oxygen-deprived) conditions. These gases can diffuse into the water column or be released to the atmosphere, influencing aquatic chemistry, greenhouse gas emissions, and water quality.

### Pollutant Fluxes

Submerged soils can act as sinks or sources for pollutants such as heavy metals, pesticides, and organic contaminants. Pollutant fluxes occur through adsorption, desorption, and microbial transformation processes in the soil-water interface. Contaminants may be transported from the soil into the water column, affecting aquatic biota and human health, or vice versa, where waterborne pollutants are deposited onto the soil surface. Understanding pollutant fluxes is crucial for managing water quality, protecting ecosystems, and mitigating environmental pollution.

### Sediment Fluxes

Sediment transport is a significant flux between submerged soils and water bodies, influencing geomorphological processes, habitat dynamics, and sediment-associated pollutant transport. Erosion and deposition of sediment particles occur due to hydrodynamic forces, such as currents, waves, and tides, as well as biological activity and human disturbances. Sediment fluxes contribute to the formation of aquatic habitats, nutrient cycling, and sedimentation-related issues, such as sedimentation of waterways and reservoirs.

### Biological Fluxes

Biological activity, including microbial metabolism, plant root exudation, and faunal bioturbation, drives fluxes of organic matter, nutrients, and gases in submerged soils. Microbial communities play essential roles in organic matter decomposition, nutrient cycling, and redox reactions, influencing soil fertility and biogeochemical processes. Plant roots facilitate nutrient uptake and oxygen release into the soil, affecting soil oxygenation and nutrient availability. Faunal organisms, such as burrowing invertebrates and aquatic insects, alter soil structure and nutrient dynamics through their activities, enhancing soil-water interactions and ecosystem functioning.

## Methane and Hydrogen Sulphide Formation

Methane ($CH_4$) and hydrogen sulphide ($H_2S$) formation in submerged soils occur primarily under anaerobic (oxygen-deprived) conditions during microbial decomposition processes.

## Methane Formation (Methanogenesis)

Methane is primarily produced through a microbial process called methanogenesis, which occurs in anaerobic environments such as submerged soils, wetlands, and sediments. Methanogenic archaea, specifically methanogens, are responsible for methane production.

Methanogens are strict anaerobes that utilize simple organic compounds, such as acetate, hydrogen, and carbon dioxide, as substrates for methane production. These microorganisms carry out a series of biochemical reactions, including acetogenesis and methanogenesis, to convert organic matter into methane and carbon dioxide.

Organic matter, such as plant debris, dead microorganisms, and sedimentary carbon, serves as the primary substrate for methanogenesis. As organic matter undergoes anaerobic decomposition, methanogens break down complex organic compounds into simpler molecules, releasing methane as a metabolic byproduct.

Factors influencing methane formation in submerged soils include temperature, pH, substrate availability, redox potential, and microbial community composition. Methane production rates tend to increase under warm, waterlogged conditions with abundant organic inputs, such as in tropical wetlands and rice paddies.

## Hydrogen Sulphide Formation (Sulphate Reduction)

Hydrogen sulphide is produced through microbial sulphide reduction, a process in which sulphide ions ($SO_4^{2-}$) are enzymatically reduced to hydrogen sulphide ($H_2S$) by sulphide -reducing bacteria (SRB) under anaerobic conditions. During sulphate reduction, sulfate-reducing bacteria enzymatically convert sulphate ions into hydrogen sulphide through a series of biochemical reactions, releasing energy for microbial growth and metabolism. Hydrogen sulphide is then released into the surrounding environment, contributing to the characteristic rotten egg odour associated with anaerobic environments. Hydrogen sulphide formation is influenced by factors such as sulphate availability, organic carbon availability, pH, temperature, and redox conditions. High organic inputs, stagnant water, and reduced sulphur compounds can enhance hydrogen sulphide production in submerged soils and sediments.

Both methane and hydrogen sulphide production in submerged soils play important roles in biogeochemical cycling, greenhouse gas emissions, and ecosystem dynamics. These gases have implications for climate change, water quality, and the functioning of aquatic ecosystems, highlighting the importance of understanding their formation mechanisms and environmental impacts.

# Salt Affected Soils

The salt-affected soils occur in the arid and semiarid regions where evapo-transpiration greatly exceeds precipitation. The accumulated ions causing salinity or alkalinity include sodium, potassium, magnesium, calcium, chlorides, carbonates and bicarbonates. The salt- affected soils can be primarily classified as saline soil and sodic soil.

## Saline Soils

Saline soils defined as soils having a conductivity of the saturation extract greater than 4 dS $m^{-1}$ and an exchangeable sodium percentage less than 15. The pH is usually less than 8.5. Formerly these soils were called white alkali soils because of surface crust of white salts.

### Formation

The process by which the saline soil formed is called Salinization. Saline soils occur mostly in arid or semi-arid regions. In arid regions saline soils occur not only because there is less rainfall available to leach and transport the salts but also because of high evaporation rates, which tend further to concentrate the salts in soils and in surface waters.

## Alkali / Sodic soils

Alkali or sodic soil is defined as a soil having a conductivity of the saturation extract less than 4 dS $m^{-1}$ and an exchangeable sodium percentage greater than 15. The pH is usually between 8.5 - 10.0. Most alkali soils, particularly in the arid and semi-arid regions, contain $CaCO_3$ in the profile in some form and constant hydrolysis of $CaCO_3$ sustains the release of $OH^-$ ions in soil solution. The $OH^-$ ions so released result in the maintenance of higher pH in calcareous alkali soils than that in non - calcareous alkali soils.

## Saline Alkali or Sodic Soils

Saline-alkali / sodic soil is defined as a soil having a conductivity of the saturation extract greater than 4 dS $m^{-1}$ and an exchangeable sodium percentage greater than 15. The pH is variable and usually above 8.5 depending on the relative amounts of exchangeable sodium and soluble salts. When soils dominated by exchangeable sodium, the pH will be more than 8.5 and when soils dominated by soluble salts, the pH will be less than 8.5.

# Acid Sulphate Soil (ASS)

Acid sulphate is drained coastal wetland soils that have become acid ($pH<4$) due to oxidation of the pyritic minerals in the soil. Undrained soils containing pyrites need not be acid and they are called potential acid sulphate soils.

## Types of Acid Sulphate Soils

### Potential acid Sulphate Soils

ASS which has not been oxidised by exposure to air are known as potential acid sulphate soils (PASS). They are neutral in pH (6.5–7.5), contain unoxidised iron sulphides, are usually soft, sticky and saturated with water and are usually gel-like muds but can include wet sands and gravels have the potential to produce acid if exposed to oxygen.

### Actual Acid Sulphate Soils

When PASS is exposed to oxygen, the iron sulphides are oxidised to produce sulfuric acid and the soil becomes strongly acidic (usually below pH 4). These soils are then called actual acid sulphate soils (AASS). They have a pH of less than 4, contain oxidised iron sulphides, vary in texture and often contain jarosite (a yellow mottle produced as a by- product of the oxidation process).

### Iron Pyrites

Iron pyrites, also known as pyrite or fool's gold, are a mineral composed of iron and sulphur ($FeS_2$). They appear as brassy yellow metallic crystals and are commonly found in sedimentary rocks, hydrothermal veins, and coal deposits. Iron pyrites form through various geological processes such as hydrothermal deposition and sedimentary accumulation. Historically valued for their resemblance to gold, they have limited industrial applications due to their high sulphur content. They are primarily mined for sulphur production and may contain trace amounts of valuable metals like gold, silver, and copper.

## Soil Reclamation

Soil reclamation is the process of restoring damaged or contaminated soils to a healthy, productive state. Techniques include soil aeration, organic amendments, phytoremediation, and erosion control. It's used in various contexts, from industrial sites to agricultural lands, to mitigate environmental damage and promote sustainability. Challenges include high costs and technical complexities, but successful reclamation restores soil functionality and supports ecosystem health.

Chapter-11

# Soil Analysis

Soil analysis is a useful tool to ensure the efficient use of applied plant nutrients. Soil tests provide a means for assessing the fertility status of a soil, but do not provide a direct measure of the actual quantity of plant nutrients in the soil. Instead, soil tests measure the quantity of a nutrient element that is extractable from a soil by a particular chemical extracting solution.

## Collection of Soil Samples

### Procedure

- Divide the field into different homogenous units based on the visual observation and farmer's experience.
- Remove the surface litter at the sampling spot.
- Drive the auger to a plough depth of 15 cm and draw the soil sample.
- Collect at least 10 to 15 samples from each sampling unit and place in a bucket or tray.
- If auger is not available, make a 'V' shaped cut to a depth of 15 cm in the sampling spot using spade.
- Remove thick slices of soil from top to bottom of exposed face of the 'V' shaped cut and place in a clean container.

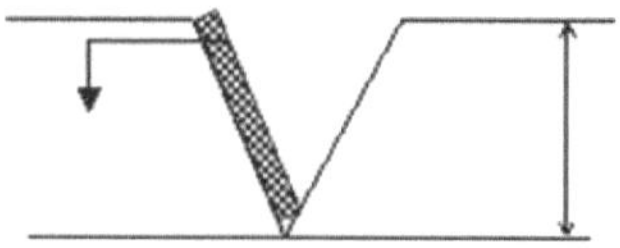

1 inch / 2.5 cm 6 inches (15 cm)

- ☐ Mix the samples thoroughly and remove foreign materials like roots, stones, pebbles and gravels.
- ☐ Reduce the bulk to about half to one kilogram by quartering or compartmentalization.
- ☐ Quartering is done by dividing the thoroughly mixed sample into four equal parts. The two opposite quarters are discarded and the remaining two quarters are remixed and the process repeated until the desired sample size is obtained.
- ☐ Compartmentalization is done by uniformly spreading the soil over a clean hard surface and dividing into smaller compartments by drawing lines along and across the length and breadth. From each compartment a pinch of soil is collected. This process is repeated till the desired quantity of sample is obtained.
- ☐ Collect the sample in a clean cloth or polythene bag.
- ☐ Label the bag with information like name of the farmer, location of the farm, survey number, date of collection, name of the sampler *etc.*

## Collection of Soil Samples from a Profile

- ☐ After the profile has been exposed, clean one face of the pit carefully with a spade and note the succession and depth of each horizon.
- ☐ Prick the surface with a knife or edge of the spade to show up structure, colour and compactness.
- ☐ Collect samples starting from the bottom most horizon first by holding a large basin at the bottom limit of the horizon while the soil above is loosened by a khurpi.
- ☐ Mix the sample and transfer to a polythene or cloth bag and label it.

## Processing and Storage

- ☆ Assign the sample number and enter it in the laboratory soil sample register.
- ☆ Dry the sample collected from the field in shade by spreading on a clean sheet of paper after breaking the large lumps, if present.
- ☆ Spread the soil on a paper or polythene sheet on a hard surface and powder the sample by breaking the clods to its ultimate soil particle using a wooden mallet.

- ☆ Sieve the soil material through 2 mm sieve.
- ☆ Repeat powdering and sieving until only materials of >2 mm (no soil or clod) is left on the sieve.
- ☆ Collect the material passing through the sieve and store in a clean glass or plastic container or polythene bag with proper labelling for laboratory analysis.
- ☆ For the determination of organic matter, it is desirable to grind a representative sub sample and sieve it through 0.2 mm sieve.
- ☆ If the samples are meant for the analysis of micronutrients atmost care is needed in handling the sample to avoid contamination of iron, zinc and copper. Brass sieves should be avoided and it is better to use stainless steel or polythene materials for collection, processing and storage of samples.
- ☆ Air-drying of soils must be avoided if the samples are to be analysed for $NO_3$-N and $NH_4$-N as well as for bacterial count.
- ☆ Field moisture content must be estimated in un-dried sample or to be preserved in a sealed polythene bag immediately after collection.
- ☆ Estimate the moisture content of sample before every analysis to express the results on dry weight basis.

## Sampling Tools

Using a soil probe, soil auger, or spade can collect surface soil samples; however, for collecting subsoil samples, a spade is not very satisfactory. A post-hole digger can be used for collecting deep samples but requires some special techniques.

## Soil Probe

- ❑ Ideal for minimal soil disturbance
- ❑ Provides continuous cores
- ❑ Suitable for dividing samples by depth
- ❑ Minimal contamination between surface and subsoil
- ❑ Unsuitable for wet, dry, or frozen soil, and gravel-containing soil

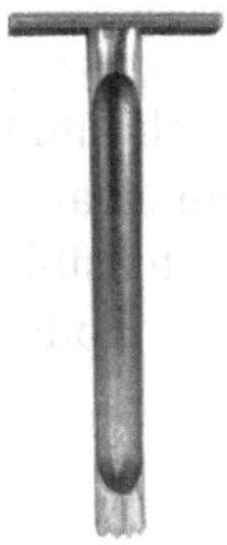

## Soil Auger

- Can be used in frozen or gravel-containing soil.
- Requires careful sampling to avoid mixing depths.
- Not suitable for wet, sticky soil or dry, powdery soil.
- Use only when a soil probe is not viable.

## Proper Sampling Depth

Surface (0-15cm depth) samples are commonly utilized for analysing soil pH, lime requirement, organic matter content, as well as levels of phosphorus, potassium, sulphur, and available micronutrients. Soil test correlations and calibrations for these parameters are predominantly established using surface samples. The tillage layer is typically defined as the 0-15cm or 0-20cm depth.

## Time of Sampling

Winter is optimal for soil sampling, except for nitrate - nitrogen testing in sandy soils. Sampling can be conducted year-round, excluding the rainy season in farmer fields. Delaying sampling until spring allows for soil moisture replenishment from fall, winter, and early spring precipitation, providing a more accurate reflection of nitrate nitrogen distribution during crop growth.

## Drying

Soil samples are placed in paper bags, cardboard boxes, or aluminum trays and dried in a rack or cabinet with exhaust fans to hasten moisture loss. This step is crucial, especially for potassium analysis, as drying temperatures can significantly affect results. For nitrate analysis, soil should be dried or frozen within 12 hours of sampling. Samples can be spread on clean surfaces and blow-dried with a fan if needed. When artificial drying isn't feasible, samples can be spread on clean surfaces like paper plates. Crushing soil clods initially helps speed up the drying time at room.

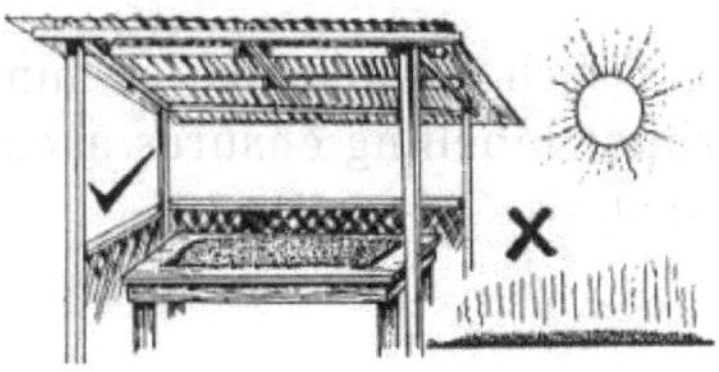

On Top of a Table Indoor Sun Drying

Room Temperature High Temperature

Air Drying Oven Drying

## Crushing and Sieving

The nature of analyses to be done, in addition to the presence of rocks or limestone concretions, determines the initial steps to crushing. Crush samples designated for mechanical analyses with a wooden rolling pin after removing all stony materials from the soil. Crush other samples with a flail-type grinder, a power-driven mortar and pestle, or some other crusher designed to minimize contamination through carryover from one sample to another. If micronutrients are to be analysed, it is important that all surfaces coming in contact with the soil be stainless steel, plastic, or wooden, preferably in the order listed. Samples should be crushed until a major portion of the sample will pass a 2 mm opening sieve. Crushing to pass a finer mesh sieve may be desirable for analysis utilizing less than one gram of soil.

### Sample Labelling

Label each sample with unique identifiers such as site name or number, sampling depth, and date of collection. Proper labelling ensures accurate identification and tracking of samples during analysis.

### Sample Storage

Store soil samples in clean, labelled containers to prevent contamination and preserve sample integrity. Keep samples cool and dry during transport to the laboratory.

## Soil Parameters

### 1. Soil Texture Analysis Methods

The soil texture refers to the relative proportions of sand, silt, and clay particles in the soil. It can be determined using various methods such as the hydrometer method, pipette method, or the feel method.

#### (i) Hydrometer Method

In this method, soil particles are separated by dispersion in water, and the sedimentation rate of particles is measured using a hydrometer. The proportions of sand, silt, and clay are determined based on the sedimentation rates.

#### (ii) Pipette Method

Similar to the hydrometer method, the sedimentation is done in a pipette. The volume of each particle size fraction settled in the pipette is measured to calculate percentages. This qualitative method involves rubbing a moist soil sample between fingers to determine the relative proportions of sand, silt, and clay based on their texture and consistency.

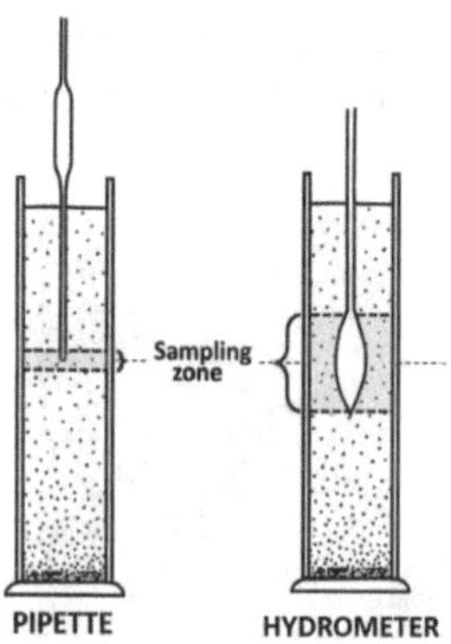


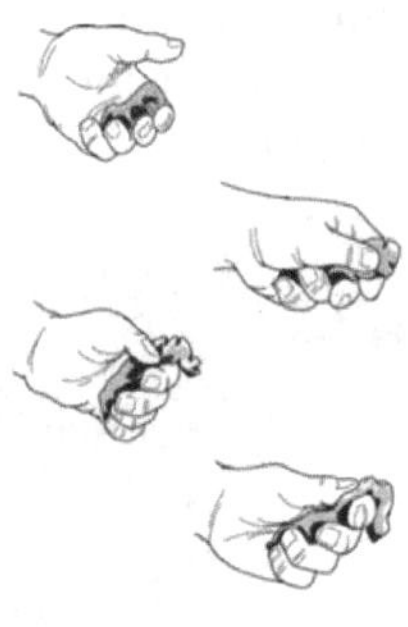

Feel Method

### 2. Water Holding Capacity

Water holding capacity (WHC) refers to the ability of soil to retain water against gravitational forces. It can be determined by saturating a soil sample, allowing it to drain, and measuring the retained water.

**Procedure**

Fill a known volume of soil into a container, saturate it with water, allow excess water to drain, and measure the water retained by the soil. WHC is expressed as the volume or mass of water retained per unit volume or mass of soil.

## 3. pH

Soil pH is a measure of acidity or alkalinity and is determined using a pH meter or pH indicator solution.

**Procedure**

Mix a soil sample with water to form a slurry, and then measure the pH of the slurry using a calibrated pH meter or add indicator solution to the soil-water mixture and compare the colour change to a standard colour chart.

**Interpretation**

Soil pH influences nutrient availability, microbial activity, and soil structure. Generally, soils with pH below 7 are acidic, while those above 7 are alkaline.

## 4. Soil Conductivity

Soil conductivity measures the ability of soil to conduct electrical current, which is related to the concentration of dissolved ions in the soil solution.

**Procedure**

Extract soil solution using a suitable method and measure its electrical conductivity (EC) using a conductivity meter.

**Interpretation**

Soil conductivity provides information about soil salinity, nutrient levels, and soil fertility. High conductivity may indicate salinity or excess fertilizer application.

## 5. Organic Carbon in Aquaculture

Organic carbon in aquaculture systems refers to the presence of carbon-containing compounds derived from organic matter such as uneaten feed, faeces, decaying plant material, and microbial biomass.

**Importance**

Organic carbon serves as a source of energy and nutrients for microbial communities, aquatic plants, and filter-feeding organisms in aquaculture systems. It plays a vital role in nutrient cycling, water quality dynamics, and overall ecosystem functioning.

## Sources of Organic Carbon

- **Feeds:** Organic carbon enters aquaculture systems through the feeding of fish, shrimp, and other cultured organisms. Uneaten feed and metabolic waste contribute to organic carbon accumulation in water and sediments.

- ☆ **Organic Waste:** Decomposition of organic matter, including faeces, dead organisms, and decaying plant material, releases organic carbon into the aquatic environment.
- ☆ **Microbial Activity:** Microorganisms play a crucial role in organic carbon cycling, breaking down complex organic compounds into simpler forms through processes such as decomposition and mineralization.

**Methods Used for Organic Carbon Analysis**

- ❑ **Walkley-Black Method:** This method involves oxidizing soil organic matter with potassium dichromate in sulfuric acid, followed by titration with ferrous ammonium sulphate. The amount of ferrous ions consumed during the reaction is proportional to the organic carbon content in the soil.
- ❑ **Loss on Ignition (LOI):** In this method, soil samples are heated to high temperatures (typically around 500-600°C) to burn off organic matter, leaving behind inorganic residues. The weight loss after ignition represents the organic carbon content of the soil.
- ❑ **Dry Combustion Method:** Soil samples are combusted in a furnace at high temperatures (550-600°C) to oxidize organic carbon into carbon dioxide. The evolved carbon dioxide is then quantified using infrared gas analyzers or coulometric titration.
- ❑ **Muffle Furnace Method:** Soil samples are combusted in a muffle furnace at high temperatures (500-600°C) to burn off organic matter. The weight loss after combustion is used to calculate the organic carbon content.
- ❑ **Elemental Analyzer Method:** This method involves the combustion of soil samples at high temperatures (around 950°C) in a pure oxygen atmosphere. The evolved gases (carbon dioxide and nitrogen oxides) are separated and quantified using gas chromatography or infrared detection.

**Effects**

- ❑ **Nutrient Recycling:** Organic carbon serves as a substrate for microbial decomposition, releasing nutrients such as nitrogen and phosphorus back into the water column for reuse by aquatic organisms.
- ❑ **Water Quality:** Excessive organic carbon accumulation can lead to degraded water quality, oxygen depletion, and the formation of harmful metabolites such as ammonia and hydrogen sulphide, which can stress or harm aquatic organisms.
- ❑ **Sediment Dynamics:** Organic carbon deposition in sediments contributes to sediment organic matter content, affecting sediment stability, nutrient cycling, and benthic community composition in aquaculture ponds or tanks.

### Management

- **Feed Management:** Optimizing feed composition, feeding rates, and feeding strategies can reduce organic carbon inputs and minimize nutrient loading in aquaculture systems.
- **Water Exchange:** Regular water exchange or circulation helps dilute organic carbon concentrations, improve water quality, and reduce the risk of oxygen depletion and nutrient accumulation.
- **Biological Filtration:** Implementing biological filtration systems, such as biofilters or vegetated wetlands, enhances organic carbon removal and nutrient assimilation, promoting water quality improvement and ecosystem balance.

## 6. Nitrogen

Nitrogen in soil exists in various forms, including organic nitrogen, ammonium ($NH_4^+$), nitrate ($NO_3^-$), and nitrite ($NO_2^-$). These forms of nitrogen play essential roles in soil fertility and plant nutrition.

- **Organic Nitrogen:** This form of nitrogen is present in organic matter, such as decaying plant and animal residues, and is not immediately available to plants. Organic nitrogen must undergo mineralization, a process in which microorganisms convert it into ammonium and subsequently into nitrate, before plants can utilize it.
- **Ammonium ($NH_4^+$):** Ammonium is a positively charged ion formed by the breakdown of organic matter or from the decomposition of urea or ammonium-containing fertilizers. It is the predominant form of nitrogen in acidic soils or under anaerobic conditions. Ammonium is readily available for plant uptake but can be converted to nitrate through nitrification.
- **Nitrate ($NO_3^-$):** Nitrate is a negatively charged ion formed through the oxidation of ammonium by nitrifying bacteria. It is the primary form of nitrogen taken up by plants and is highly mobile in soils, making it susceptible to leaching below the root zone. Excessive nitrate leaching can lead to groundwater contamination and environmental issues.
- **Nitrite ($NO_2^-$):** Nitrite is an intermediate product formed during the nitrification process, where ammonium is oxidized to nitrate. It is less common in soils and is rapidly converted to nitrate by nitrifying bacteria.

### Methods Used for Nitrogen Analysis

- **Kjeldahl Method**: This method involves digesting the soil sample with concentrated sulfuric acid to convert nitrogen into ammonium ions, followed by distillation and titration to quantify the ammonium ions.
- **Total Kjeldahl Nitrogen (TKN)**: Similar to the Kjeldahl method, but it includes additional steps to convert organic nitrogen compounds into ammonium ions before distillation and titration.

- **Dumas Combustion Method**: This method combusts the soil sample at high temperatures, converting nitrogen into nitrogen gas, which is then measured using a detector.
- **Micro-Kjeldahl Method**: A modified version of the Kjeldahl method suitable for smaller sample sizes.

**Effects**

- **Plant Growth and Development:** Nitrogen is a primary component of amino acids, proteins, nucleic acids, and chlorophyll, essential for plant growth and development. Adequate nitrogen availability in soil promotes vigorous vegetative growth, increased leaf area, and improved photosynthetic efficiency, leading to higher crop yields and improved crop quality.
- **Nutrient Cycling:** Nitrogen is a key nutrient in the soil nutrient cycling process. It undergoes various transformations, including mineralization (conversion of organic nitrogen to ammonium), nitrification (conversion of ammonium to nitrate), and denitrification (conversion of nitrate to nitrogen gas). These processes are carried out by soil microorganisms and are essential for recycling nitrogen within ecosystems.
- **Soil Fertility:** Nitrogen availability in soil directly influences soil fertility. Soils with adequate nitrogen levels support healthy plant growth and maintain soil productivity over time. However, nitrogen deficiency can lead to stunted growth, reduced yields, and nutrient imbalances in plants. Conversely, excessive nitrogen can lead to nutrient imbalances, environmental pollution, and ecosystem degradation.
- **Microbial Activity:** Nitrogen availability influences soil microbial communities and their activities. Soil bacteria and fungi play crucial roles in nitrogen cycling processes, such as nitrogen fixation by nitrogen-fixing bacteria, nitrification, and denitrification. Changes in soil nitrogen levels can affect the composition and function of soil microbial communities, which, in turn, can influence nutrient cycling and ecosystem processes.
- **Soil pH and Acidity:** Nitrogen fertilizers, such as ammonium-based fertilizers, can lower soil pH and increase soil acidity over time due to nitrification processes. Acidification of soils can affect nutrient availability, soil structure, and microbial activity, leading to soil degradation and reduced crop productivity in the long term.
- **Environmental Impacts:** Excessive nitrogen application can lead to environmental issues such as nitrate leaching, groundwater contamination, and emissions of nitrous oxide, a potent greenhouse gas. Nitrate leaching can pollute surface and groundwater, posing risks to human health and aquatic ecosystems. Nitrous oxide emissions contribute to global warming and climate change.

### Management

- **Soil Testing and Monitoring:** Regular soil testing allows for accurate assessment of nitrogen levels and helps in making informed decisions regarding fertilizer application rates and timing.
- **Appropriate Fertilizer Application:** Applying fertilizers at the correct rate, timing, and placement ensures efficient utilization of nitrogen by crops while minimizing losses through leaching, volatilization, and runoff.
- **Crop Rotation and Cover Cropping:** Implementing diverse crop rotations and incorporating cover crops, especially nitrogen-fixing legumes, helps maintain soil nitrogen levels, improves soil health, and reduces the need for synthetic fertilizers.

## 7. Phosphorous

Phosphorus is an essential nutrient for aquatic organisms, playing a vital role in bone formation, metabolic processes, and energy transfer.

### Sources

- Commercial aquafeeds contain phosphorus in various forms to meet the nutritional requirements of cultured species.
- Decomposition of organic matter, including uneaten feed and faeces, releases phosphorus into the aquatic environment.
- Phosphorus-based fertilizers may be used in some aquaculture systems to enhance primary productivity and promote the growth of phytoplankton and other food sources.

## Methods for Analysis of Phosphorous in Soil

There are several methods commonly used for the analysis of phosphorus in soil.

#### Colorimetric Methods

- **Molybdate-Blue Method:** This method involves the reaction of phosphorus with ammonium molybdate and antimony potassium tartrate under acidic conditions to form a blue complex. The intensity of the blue colour is measured spectrophotometrically at a specific wavelength, and the phosphorus concentration is determined by comparing it to a calibration curve.
- **Vanadomolybdate Method:** Similar to the molybdate-blue method, this method also involves the formation of a coloured complex between phosphorus and ammonium molybdate. However, vanadium (V) is used as the reducing agent instead of antimony, resulting in a yellow colour. The intensity of the yellow colour is measured spectrophotometrically for phosphorus quantification.

- **Ion-Selective Electrode (ISE) Method:** This method utilizes a specific ion-selective electrode for phosphorus detection. The electrode responds selectively to phosphate ions in the soil extract, generating an electrical signal proportional to the phosphorus concentration. Calibration curves are used to relate the electrical signal to phosphorus concentration.
- **Sequential Extraction Methods:** These methods involve sequential extraction of phosphorus from soil using different chemical reagents to fractionate phosphorus into different forms based on their availability to plants. Common extraction reagents include Mehlich-3, Bray P-1, and Olsen solution. Each extraction step targets specific phosphorus pools, such as labile, moderately labile, and non-labile phosphorus fractions. The extracted phosphorus fractions are then analysed using colorimetric or ISE methods.

**Effects**

- Adequate phosphorus levels are essential for fish and other aquatic organisms' growth, bone development, and reproductive functions.
- Excessive phosphorus can contribute to eutrophication, algal blooms, and degraded water quality, posing risks to aquatic ecosystems and aquaculture production.

**Management Practices**

- Monitoring and managing phosphorus levels along with other nutrients to maintain a balanced nutrient profile in aquaculture systems.
- Optimizing feed composition and feeding practices to minimize phosphorus waste and nutrient loading in water bodies.
- Regular monitoring of phosphorus concentrations and water quality parameters to assess system health and prevent eutrophication.

## Liming

Liming is considered one of the most crucial steps in pond preparation and treatment, significantly contributing to water quality improvement in many cases. Lime serves as an effective conditioner in ponds, particularly when applied after ploughing to achieve a clear and smooth pond surface. Additionally, lime plays a vital role in enhancing pond sanitation and increasing fish pond productivity. Promoting the healthy growth of fish in the pond relies heavily on the presence of phytoplankton, whose growth and survival are influenced by various nutrient components in the pond soil. Lime application is thus essential for rendering these nutrient components usable, ensuring optimal conditions for phytoplankton and, consequently, fish growth and health.

### Functions of Lime in the Fish Pond

The main function of the lime in the fish pond is to reduce the acidity of the soil and water. Other accessory functions of lime in the fish pond are:

- To increase the natural food in the pond, this is essential for the better growth and survival of fish. Lime accelerates the process of degradation of organic matter in the pond, so that amount of usable nitrogen in the pond increased rapidly. As a result, natural food for fish grows adequately and sufficiently in the pond.
- To increase the amount of usable phosphate in the pond lime is necessary for the growth of phytoplankton. The amount of usable phosphate in the fish pond largely depends upon pH of the soil. If this pH of the soil is in the range of 6.5-7.5, then the amount of usable phosphate in the pond is good. In the case of excessive acidic water, especially when the pH of the soil, is lower than 6.5, the amount of usable phosphate in the pond becomes very low. The main reason for this phenomenon is that most of the phosphate combines with iron and aluminum, so plant usable phosphate becomes extremely low. Application of lime in the fish pond, reduce the acidity of the soil of the fish pond. As a result of that, phosphate becomes free and become usable by plants.
- The use of lime in ponds raises water alkalinity, serving as a buffer to maintain optimal pH levels throughout the day, fostering ideal conditions for fish culture. As lime contains calcium, so this increases the productivity of the pond through the application of lime.
- Lime helps to cure some fish diseases through its extra alkaline characteristics.
- Lime increases the oxygen content of the pond water by changing physical conditions of the water.
- Lime increases the transparency of the water.
- Lime decreases the amount of toxic gases in the pond.

## The Optimum Amount of Lime Used in the Pond

Amount of the lime to neutralize the pond water largely depends upon the pH of the soil and water of the pond. If the acidity of the pond bottom soil increases, the amount of lime for neutralization also increases. If pH of the pond is neutral, then the amount of lime required also decreases.

**Proper application time of lime:** Optimum time for application of lime in the water is early morning as water pH becomes very low at this time.

## Types of Lime and their Effectiveness

### Calcium Oxide or Quick Lime

- **Chemical formula:** CaO
- **Nature:** This lime produces heat when it mixes with water. This lime very rapidly increases the pH of soil and water.

- **Effectiveness:** This lime is very much effective in those ponds which are highly acidic. This lime is very much effective in the prevention of fish diseases.

## Calcium Hydroxide or Hydrated Lime

- **Chemical formula: $Ca(OH)_2$**
- **Nature:** Acidity neutralizing power is less than that of quick lime. This lime produces less heat than quick lime.
- **Effectiveness:** Use of this lime in the pond is very less.

## Calcium Carbonate

- **Chemical formula: $CaCO_3$**
- **Nature:** Acid neutralizing power is very much less than that of other lime. This lime does not produce heat when it mixes with water.
- **Effectiveness:** As this lime does not increase the pH of the pond rapidly, so it is suitable to apply in the alkaline pond where pH of the pond water varies between 8-8.5. Other benefits of the application of lime are also extracted from the lime.

## Dolomite

- **Chemical formula: $CaMg(CO_3)_2$**
- **Nature:** This lime has very low acid-neutralizing power and contains magnesium in addition to calcium.
- **Effectiveness:** It increases the magnesium content in water, which is essential for plant nutrition. While its effectiveness in freshwater is limited, combining it with other types of lime can still provide benefits.

## Chapter-12

# Soil and Water Amendments

## Soil Amendment

- Soil amendments are organic or inorganic matter added intentionally to the soil to improve texture, water retention, drainage or aeration.
- In case of sandy or rocky soil amendments are required to improve the texture and add water retention properties.
- Clay soils require the addition of material to improve texture, aeration and drainage.

## Common Inorganic Amendments

| Amendment | Function/purpose |
|---|---|
| Sand or Soil profile conditioner | To improve aeration and drainage |
| Lime or Sulphur | To raise or lower pH |
| Manures | To add nutrients |
| Mulch | Mulch helps soil retain its moisture |
| Compost | Improves the texture of the soil, increases aeration, promotes both water retention and drainage as well as adds valuable nutrients necessary for plant growth |
| Humus | Humus offers stability to imbalanced soil |
| Peat Moss | It increases the ability of the soil to retain water and is ideal for sandy or rocky soil. It is ideal for stabilizing clay soils as well |

## Advantages

- Soil amendments improve the physical properties resulting in better conditions for water storage, root development and soil ecosystems
- Soil amendments enhances soil aeration
- Soil amendments can be produced locally, especially organic amendments (e.g. compost), which are cheap to produce

## Disadvantages

- Wood products can tie up nitrogen in the soil
- The incorporation of the soil amendment (especially of organic amendments) can be time consuming
- If too many nutrients or organic matter are put into the soil, they can be released and cause an outflow of nutrients into the groundwater and surrounding rivers and lakes, which can result in water pollution

## Factors to be Considered in Selecting in Soil Amendment

- Soil texture
- Salt contents
- Soil salinity and plant sensitivity to salts
- Salt content and pH of the amendment

## Common Materials Used as Soil Amendment

- Lime, manures, fertilizers, micronutrients, zeolites, alum, gypsum.
- Lime is a calcium-containing inorganic mineral composed primarily of oxides, and hydroxide, usually calcium oxide and/ or calcium hydroxide.

## Types of Lime

- ☆ Agricultural lime ($CaCO_3$)
- ☆ Quick lime ($CaO$)
- ☆ Slaked lime $Ca(OH)_2$

## Liming Materials

- They are substances containing Ca and Mg compounds capable of neutralizing acidity.
- The most common liming materials are oxides, hydroxides and carbonates of Ca and Mg having higher percentage of Ca compared to Mg.

Chemical changes due to different liming materials, which occur, may be represented as follows:

$$CaO + 2H^+ \rightarrow Ca^{2+} + H_2O$$

$$Ca(OH)_2 + 2H^+ \rightarrow Ca^{2+} + 2H_2O$$

$$CaCO_3 + 2H^+ \rightarrow Ca^{2+} + H_2O + CO_2$$

$$CaMg\,(CO_3)^{2+}\,4H^+ \rightarrow Ca^{2+} + Mg^{2+} + 2H_2O + 2CO_2$$

- The effectiveness of liming material depends on their neutralizing capacity, particle size and physical properties.
- Pure $CaCO_3$ can be considered to have a neutralizing value of 100%, and is used as the standard.

## Benefits of Liming

- Liming establishes a strong pH buffer system, and thereby prevent sudden fluctuation in environment, that is considered detrimental for aquatic life.
- Lime reduces turbidity caused by suspended silt and clay particles and humic substances.
- Toxic and caustic actions of liming materials kill bacteria as well as fish parasite in their various stages of life cycle and thus render fishes less liable to disease.
- The pH of acidic soil raised by liming increases microbial activity, and thus favours mineralization of nitrogen and other nutrients through decomposition of organic matters.

## Methods of Application of Liming Material

- Agricultural lime is generally used in fish ponds.
- It is low priced and does not create any harmful effects, compared to oxides and hydroxides, which are used under specific conditions for control of diseases.
- Liming should be applied on to pond bed when it is desired to improve pond bottom or to control parasites.
- But for controlling precipitation of water and removal of turbidity in acid water, liming materials are broadcasted on the water surface.
- Newly constructed ponds should be limed prior to filling with water.
- Maintenance of desired pH of soil is rather difficult by liming due to relationship between reserve and exchange acidity of soil, so frequent applications of lime in split doses are considered effective to achieve desired results.

## Lime doses at different pH level

| pH | Soil condition | Dose (kg/ha) |
|---|---|---|
| 4.0- 4.5 | Highly acidic | 1000 |
| 4.5- 5.5 | Medium acidic | 700 |
| 5.5- 6.5 | Slightly acidic | 500 |
| 6.5- 7.5 | Near neutral | 200 |

# Manures

- Manure is organic matter that is used as organic fertilizer in agriculture.
- Most manure consists of animal faeces; other sources include compost and green manure.
- Manures contribute to the fertility of soil by adding organic matter and nutrients, such as nitrogen, that are utilised by bacteria, fungi and other organisms in the soil.

## Types of Manure

### Animal manure

- Cattle manure is a good source of nitrogen as well as organic carbon.
- Chicken litter, is very concentrated in nitrogen and phosphate and is prized for both properties.
- It can improve the soil structure (aggregation) so that the soil holds more nutrients and water, and therefore becomes more fertile.

### Nutritive Value of Different Manure

| Animal | Excreta | Moisture (%) | Organic matter (%) | Nitrogen (%) | Phosphorous (%) | Potash (%) |
|---|---|---|---|---|---|---|
| Cattle | Faeces | 80 – 85 | 14.0 | 0.3 | 0.2 | 0.1 |
| | Urine | 92- 95 | 2.3 | 1.0 | 0.1 | 1.4 |
| Pig | Faeces | 85 | 15.0 | 0.6 | 0.5 | 0.4 |
| | Urine | 97 | 2.5 | 0.4 | 0.1 | 0.7 |
| Chicken | Faeces | 78 | 25.5 | 1.4 | 0.8 | 0.6 |
| Duck | Faeces | 81 | 26.2 | 0.9 | 0.4 | 0.6 |
| Rabbit | Faeces | 10 | 37.0 | 2.0 | 1.3 | 1.2 |
| Goat | Faeces | 10 | - | 2.7 | 1.7 | 2.9 |

### Compost

- Composting is the natural process of recycling organic matter, such as leaves and food scraps, into a valuable fertilizer that can enrich the soil.
- Anything that grows decomposes eventually; composting simply speeds up the process by providing an ideal environment for bacteria, fungi, and other decomposing organisms (such as worms, sowbugs, and nematodes) to do their work.
- The resulting decomposed matter, which often ends up looking like fertile garden soil, is called compost.

## Green Manure

- Green manure is a term used to describe specific plant or crop varieties that are grown and turned into the soil to improve its overall quality.
- When these plants are incorporated into the soil, it breaks down, eventually releasing important nutrients, such as nitrogen.
- There are many commonly used fertilizers produced for a variety of applications. Fertilizer manufacturers are required to list the grade on each fertilizer container by the percent of nitrogen (N), phosphorus (P) as phosphoric acid ($P_2O_5$) and potassium (K) as potassium monoxide ($K_2O$).
- Additions of phosphorus in ponds usually provide a much greater increase in fish production than from either nitrogen or potassium.
- Liquid fertilizers are generally superior to the traditional granular form fertilizers in promoting rapid growth of plankton algae in farm ponds. This is because the nutrients are immediately in solution, making them more rapidly available to the algae.
- Also, smaller applications of liquid fertilizers (more as concentrates) can be used which may reduce the cost and labour of application but still improve the effectiveness of pond fertilization.

## Organic Fertilizers

- Natural fertilizers derived from plants and animals are known as organic fertilizers.
- Organic fertilizers boost the amount of organic matter in the soil, encourage microbial reproduction, and alter the physical and chemical composition of the soil.
- It is as one of the essential elements for providing green food.

Organic fertilizers can be obtained from the following products:

- Agricultural Waste
- Livestock Manure
- Industrial Waste
- Municipal Sludge

## Inorganic Fertilizers

- Chemical fertilizers generated by chemical techniques that contain nutrients for crop growth are known as inorganic fertilizers.
- The inorganic fertilizers are of the following types

### *(i) Nitrogen Fertilizers*

- Nitrogen fertilizers contain nitrogen necessary for the development of crops.
- Nitrogen, a key constituent of chlorophyll, helps main balance in the process of photosynthesis.
- It is also a part of amino acids in plants and contains protein.
- Nitrogen fertilizers improve the production and quality of agricultural products.

### *(ii) Phosphorus Fertilizer*

- Phosphorus is the principal nutrient.
- The success of a fertilizer is influenced by factors such as the effective phosphorus concentration, fertilization techniques, soil characteristics, and crop strains.
- The protoplasm of the cell contains phosphorus, which is crucial for cell growth and proliferation.
- The growth of the plants roots is aided by the phosphorus fertilizer.

## Difference between Manure and Fertilizer

| Aspect | Manure | Fertilizer |
|---|---|---|
| Source | Organic waste from animals or decomposed plants | Chemical or natural substances synthesized industrially |

| Aspect | Manure | Fertilizer |
|---|---|---|
| Composition | Contains organic matter, nutrients, and microbes | Contains concentrated nutrients in specific ratios |
| Physical Properties | Variable texture, may contain bedding material | Typically granular, powder, or liquid form |
| Chemical Properties | Contains nitrogen, phosphorus, potassium, etc. | Contains nitrogen, phosphorus, potassium, etc. |
| Biological Properties | Supports soil microbial activity | May have synthetic compounds, less microbial activity |
| Nutrient Content | Lower nutrient concentration | Higher nutrient concentration |
| Release Rate | Nutrients released slowly over time | Nutrients released quickly upon application |
| Environmental Impact | May improve soil structure and microbial activity | Risk of leaching into waterways, soil degradation |
| Application Frequency | Applied less frequently | Applied more frequently |
| Uses | Soil amendment, organic fertilizer, composting | Quick nutrient boost, precise nutrient application |
| Examples | Cow dung, compost, poultry litter | Ammonium nitrate, urea, triple superphosphate |

## Micronutrients

Micronutrients are compounds that are only required in minimal amounts which aid in the growth, development, and maintenance of the body. Zinc, cobalt, manganese and copper play a significant role in aquaculture.

### Zinc

- ❑ Zinc can be highly beneficial for fish growth mainly of silver carp and *Catla catla*.
- ❑ The recommended dose for zinc is as follows:
  - ☆ 30 kg zinc sulphate/ha/yr for soil having 0.5 ppm available zinc
  - ☆ 20 kg zinc sulphate/ha/yr for soil having 0.5 - 0.75 ppm available zinc
  - ☆ 10 kg zinc sulphate/ha/yr for soil having 0.75-1.00 ppm available zinc
- ❑ Zinc is required for the synthesis of DNA and RNA, and is therefore, essential for cell proliferation.
- ❑ For fish, zinc is second in quantitative importance, next to iron.
- ❑ At the same time excessive environmental zinc can have severe impacts upon the survival and growth of aquatic organisms.

### Cobalt

- ❑ It plays an important role in nitrogen fixation by rhizobium and it is also a constituent of Vit $B_{12}$.

- 10 kg cobalt chloride/year as micronutrient fertilizer could be highly beneficial for fish growth especially of silver carp and Mrigala.

## Manganese

- It plays an important role in photosynthesis and heterotrophic growth of plankton.
- Application of manganese increase plankton to biomass by about 15%.
- Manganese concentration in natural surface water seldom reaches 1.0 mg/l and is usually less than 0.2mg/l.
- Higher concentration of manganese is toxic to aquatic organism.

## Copper

- Copper is mainly applied to aquacultural ponds to control algal bloom, to kill organisms, which produces odorous compounds responsible for off flavour in fish and shrimp and to control fish disease.
- The maximum admissible copper concentration in water for the protection of fish is in the range of 0.001 -0.02 mg/l depending upon the physical and chemical properties of the water and the species of the fish.
- Excess copper concentration of water is toxic to the aquatic species effecting bronchial function and growth inhibition.

## Zeolite

- Zeolite is a naturally occurring rock that has a fairly unique structure which is highly porous whilst also allowing for rapid uptake and loosing of charged particles.
- When placed in water zeolite filters ammonia out of the water and holds it inside its' porous structure.
- As ammonia is highly toxic to fish this is a very useful way of reducing ammonia concentration in water.
- Zeolite is often used when transporting fish for a period of several hours to avoid the ammonia accumulation in the bags or transport tank from reaching toxic levels.
- It can also be used when setting up a new recirculating system or tank to maintain low ammonia levels while the biofilter is maturing.
- Occasionally zeolite is used in reef aquarium to polish the water of unwanted particles other than ammonia.

## Alum

- Alum is the most effective for clearing clay turbidity from a pond.
- It can be sprinkled as a powder over the water surface, through a waterfall, or through a skimmer intake.

- Alum also reduces dissolved inorganic phosphate levels through the precipitation of insoluble aluminium phosphate, reducing phytoplankton growth.
- The use of alum might require adding lime to adjust total alkalinity and pH if these parameters drop below acceptable levels.

## Gypsum

- It is used to increase the total hardness of waters beyond that possible with lime, although it does not neutralise acidity.
- The low cost and high solubility of gypsum makes it ideal for use in the maintenance of calcium levels in hatchery situations.

Chapter-13

# Environmental Ameliorative

## Chlorination

- Chlorine disinfection is a point-of-entry treatment that kills pathogens, including certain viruses and bacteria that retard the growth of fishes. Moreover, chlorination can offer residual disinfection throughout the household water distribution system.
- Hypochlorous acid and hypochlorite (free chlorine residual) are responsible for disinfecting power of chlorine products in pond water.
- But, chlorination of waters containing fish or prawn is dangerous and unbeneficial.
- It is possible to disinfect bottoms of empty ponds and water in newly filled but unstocked ponds, by applying chlorine products.
- When this is done, enough chlorine should be applied to overcome chlorine demand to provide 1 mg/l or more of free chlorine residuals. The residuals will detoxify naturally in few days so to stock ponds safely.

## How Chlorination Works?

- The effectiveness of chlorination depends on various factors, including water temperature, water pH, water turbidity, general water quality and contact time.
- The contact time is the time available to complete the reaction between the chlorine and untreated water. A longer contact time results in more effective disinfection.

- As the chlorine concentration increases, the required contact time decreases. Chlorination is more effective at a high temperature and a low pH.
- Particles in the water decrease the effectiveness of chlorination, as microorganisms may 'hide' behind particles and avoid disinfection.
- The quality of the untreated water affects the chlorine demand. Chlorine readily combines with other components dissolved in water, including iron, manganese, hydrogen sulphide, microorganisms, plant material, ammonia, and organic colour such as that from decaying peat moss. These things 'use up' chlorine so it is important to add sufficient chlorine to meet the demand and still provide residual disinfection.
- Chlorine is available in two formulations, as a dry powder or pellet (calcium hypochlorite), or as a liquid (sodium hypochlorite).
- Chlorine solutions from powders should be prepared frequently since the strength of the solution decreases gradually after mixing. To avoid hardness deposits on equipment, manufacturers recommend using softened or distilled water when mixing chlorine solutions.

## Deodorizer

- In a confined aquaculture pond, gradual accumulation of unutilized feed, dead algae, faecal matter etc. raises the organic load at the pond bottom.
- This load absorbs all the available oxygen at pond bottom, by way of oxidation process and creates anaerobic condition.
- In the absence of oxygen, noxious gases like hydrogen sulphide ($H_2S$), ammonia ($NH_3$), nitrite ($NO_2^-$) etc. become more stable and these depleted dissolved oxygen conditions trigger stress to the cultured species and cause severe health problems ultimately leads to mortality if timely action is not taken. Few deodorizers used in aquaculture are;

## Tetra San

It is a versatile and powerful water sanitizer for use as an effective disinfectant in aquaculture ponds. It improves the total aquatic environment. By way of reducing the load of pathogenic flora and fauna in aquaculture facilities (farms and hatcheries) and also as a deodorizer for use in aquaculture ponds.

## Yuka G

Yuka G contains *Yucca schidigera* blended with suitable probiotics and enzymes in granular form.

## Benefits

- Reduces ammonia and other irritant gases
- Reduces waste odour

- Improves sludge characteristics for better
- Overall improvement in pond environment
- Easy application

## Dosage

- **Shrimp ponds:** 500 gms/ha
- **Fish/other species ponds:** 500 gms/ha
- **Packaging:** 500 gm bottle

## Common Causes of Smelly Pond Water & Odours:

- Stagnant Pond Water (Low aeration)
- Poor Water Filtration
- Pond Sludge (Bottom muck)
- Overgrowth of Algae
- Dead pond fish or plants

## Bacterial Formulation

Microbial formulations have the potential to restore the fertility of degraded soil through diverse processes, namely, by increasing nutrient availability through nitrogen fixation and mobilization of essential elements (phosphorus, iron, and potassium) to the plants, and repairing the soil structure by improving its aggregation and stability. Microbial formulations also help in disease management. The success rate of these formulations in the field depends on their antagonistic or synergistic interaction with indigenous microbes, organic carbon content, soil properties (physical and chemical), and soil amendments. A consortium of microorganisms, with or without organic fertilizer, is more beneficial for restoring soil fertility and organic matter content in agricultural soil than a single inoculum. Agricultural practices also play a significant role in the establishment of these microbial formulations.

Chapter-14

# Impact of Soil Quality on Aquaculture

Soil is one of the important factors that determine the productivity of a pond. The production of various primary food organisms depends largely on the availability of different nutrients. The availability of most nutrient in an aquaculture pond depends mostly on the prevailing condition of the bottom soil, for which it is sometimes called **"Chemical Laboratory of the pond".**

Some criteria or requirement for aquaculture are as follows:

## Texture

Nature and properties of the parent material forming soil determine soil texture. Many important physico-chemical properties for fertility of fish ponds are influenced to a great extent by the proportion of different size fraction of soil. An ideal pond should not be too sandy to allow leaching nor be too clayey to keep all the nutrients absorbed. Sandy clay, sandy clay loam, and clay loam are some of the textured soils that are suitable for aquaculture.

## Soil Acidity

Soil can either be acidic, alkaline or neutral. pH range of 6-8 is considered ideal. Soil which are acidic does not respond well to fertilization. Liming is the only way to improve water quality in ponds with acidic soil and it is the pH of the soil that must be corrected for lasting effects rather than pH of water.

## Bottom-Soil Oxidation

Dissolved oxygen cannot move rapidly into water-saturated soil, and pond soils become anaerobic below the depth of a few millimetres. Aeration and water circulation are beneficial in improving bottom-soil oxidation, but surface layer of the soil may still become anaerobic in intensive fish culture ponds. When redox potential is low at the soil surface (anaerobic condition), hydrogen sulphide and other toxic microbial metabolites diffuse into pond water. Sodium nitrate can serve as a source of oxygen for microbes is poorly oxygenated environments; the redox-potential will not drop enough for formation of hydrogen sulphide and other toxic metabolites.

## Drying Pond Bottoms

When pond bottoms are dried between crops, evaporation of water from soil-pores and cracking of soil enhances aeration and favours microbial decomposition of soil organic matter. Excessive drying makes soil too dry for even microbial activity; a drying period of 2-3 weeks is usually adequate. Tilling of dry soil with a disk harrow can improve aeration, but tilled bottoms of aerated ponds should be compacted before refilling to reduce tendency of erosion.

## Good Practices for Bottom Soil Management

Major concerns in pond bottom soil-management are low soil pH, high soil-organic matter, loss of oxidized layer, and accumulation of soft sediments. Some good practices for protecting soil quality are shown in Table.

### Good management practices for pond bottom-soil quality

| Problem | Preventive measures |
|---|---|
| Low soil pH | • Neutralize acidity of new pond bottoms - soil before initiating aquaculture.<br>• In old ponds that have never been limed, apply agricultural limestone according to soil pH.<br>• Use urea and ammonium fertilizers conservatively as they are acid-forming.<br>• Monitor total alkalinity of pond waters and soil pH to ensure that total alkalinity is >75 ppm in fish ponds and pH is >7.<br>• After initial correction of soil pH, apply agricultural limestone to bottoms of empty ponds @1000 kg/ha during fallow periods between crops. If ponds are full, apply to water. |

| Problem | Preventive measures |
|---|---|
| High soil organic matter | • Select sites without organic soil<br>• Where soils are organic, apply agricultural limestone and urea (200-400kg/ha) to encourage degradation of organic matter during fallow periods. Repeat after each crop.<br>• Use moderate stocking rates to avoid high inputs of nutrients and organic matter in fertilizers, manures and feeds.<br>• Dry ponds between crops, apply agricultural limestone according to soil pH and till heavy-textured soils to encourage oxidation of organic matter by bacteria.<br>• In areas, where pond bottom soils cannot be dried, apply nitrate fertilizer at 20 to 40 $g/m^2$.<br>• Monitor soil organic matter concentrations annually. >3% organic carbon (about 6% organic matter) suggests excessive organic matter. |
| Loss of oxidized layer | • Where a surface layer high in organic matter has developed in bottom soils, use turning plough to bury this layer and expose higher quality soil.<br>• Monitor appearance of soil. The upper few millilitres should be of natural soil colour or brownish, A gray or black colour at surface indicates reduced (anaerobic) conditions.<br>• Use a rake or chain to scarify surface soil during crop if it appears anaerobic<br>• Remove accumulation of organic matter from corners of ponds. Maintain adequate plankton to restrict light and prevent mats of benthic algae.<br>• Remove soft sediment from ponds. |
| Excessive accumulation of soft sediment | • If water supply has high concentrations of suspended solids, pass water through a settling basin before putting it in ponds.<br>• Establish grass cover to minimise erosion on watersheds and embankments of ponds.<br>• Use proper side slopes and compaction when constructing new ponds or renovating old ones.<br>• In ponds with mechanical aeration, install aerators to prevent water currents from eroding insides of embankments.<br>• If sites of active erosion are observed, measures for lessening erosion should be installed. These measures may include installation of rip-rap, proper sloping and compaction, grass cover, etc.<br>• When ponds are empty between crops, remove sediments from deep areas and place them on areas from which eroded. Proper sloping and compaction, establishment of grass above water level, or installation of rip-rap will lessen erosion potential. |

| Problem | Preventive measures |
|---|---|
| | • Do not leave ponds empty longer than necessary during rainy season to prevent erosion of soil from shallow area with deposition of soil in deeper areas.<br>• If bottom of heavily aerated ponds are tilled between crops, compact bottoms with heavy roller before refilling. Do not allow livestock to walk on pond embankments or wade in shallow water-edges.<br>• Avoid operating equipment that will cause ruts and other inundations in pond bottoms.<br>• Monitor pond bottoms for soft sediment and remove such sediments periodically instead of waiti ng until a severe problem develops. |

## Chemical Properties of Bottom Soil

Some important chemical properties of soil in pond that influence nutrient management of ponds are:

**Soil reaction (pH):** pH is one important factor that influence the productivity of a pond as it controls most of the chemical reactions in the pond. pH of near neutral and slightly alkaline is considered ideal for fish production.

**Organic carbon content:** Organic carbon acts as a source of energy for bacteria and other microbes that release nutrient through various biochemical processes. Pond soils with < 0.5% organic carbon are considered unproductive while those with 0.5-1.5% and 1.5-2.5% have medium and high productivity respectively, while > 2.5% is also not suitable for fish production as it may lead to eutrophication and oxygen depletion.

**Carbon: Nitrogen ratio:** This ratio influences activity of soil microbes which in turn affects rate of release of nutrient from decomposing organic matter. Rate of mineralisation is very fast, moderately fast and slow at C:N ratio range of <10, 10-20 and more than 20 respectively. Carbon nitrogen ratio of between 10-15 is considered favourable for aquaculture.

**General nutrient status:** Nitrogen, phosphorous and potassium are major nutrients required by phytoplankton. The single most critical nutrient of pond productivity is phosphorous content of soil and water. Pond soil with 30ppm, 30-60 ppm and 60-120 ppm and more than 120 ppm available phosphate are considered poor, average and good and highly productive respectively. Nitrogen content of <250 ppm is considered low in productivity while concentration range of 250-500ppm and above 500 ppm are considered medium and highly productive.

# Glossaries

**Acid sulphate soil:** Acid sulphate soil is a name given to soils or sediments containing iron sulphides.

**Acid:** An acid is a chemical species that donates proton or hydrogen ion and/or accepts electrons.

**Alkali soils:** These are clay soils with high pH (> 8.5), a poor soil structure and a low infiltration capacity.

**Asphyxiation:** Stress in fish due to low level of dissolved oxygen in an aquatic ecosystem.

**Base:** A base is a chemical species that donates electrons, accepts protons, or release hydroxide ($OH^-$) ions in aqueous solution.

**Buffer:** Buffers are aqueous systems that tend to resist changes in pH when small amounts of acid ($H^+$) or base ($OH^-$) are added.

**Chemical properties**: Chemical properties of water generally indicate the chemical reactions displayed by water.

**Coldwater fish species**: Fish species whose congenial temperature range is generally <20$^0$C are referred to as coldwater fish species.

**Ectotherm:** Any animal whose regulation of body temperature depends on external sources, such as sunlight or a heated rock surface.

**Endotherm:** These are those animals that maintain a constant body temperature independent of the environment.

**Environmental ameliorative:** These are substances use to repair environmental problems of soil.

**Euryhaline:** Species that can tolerate wide range of salinity.

**Formality**: It is defined as the number of formula mass of any solute dissolved in one litre of solution. A solution is said to be one formal (1 F) if it contains one formula mass of the substance in one litre of solution.

**Geologic time**: The geologic time scale (GTS) is a system of chronological dating that classifies geological strata (stratigraphy) in time. It is used by geologists, palaeontologists, and other Earth scientists to describe the timing and relationships of events in geologic history.

**Gravimetry:** It is a technique through which the amount of an analyte (the ion being analysed) can be determined through the measurement of mass. Gravimetric analyses depend on comparing the masses of two compounds containing the analyte. The principle behind gravimetric analysis is that the mass of an ion in a pure compound can be determined and then used to find the mass percent of the same ion in a known quantity of an impure compound.

**Homeostasis:** It is a self-regulating process by which an organism tends to maintain stability while adjusting to conditions that are best for its survival.

**Humus:** The organic component of soil, formed by the decomposition of leaves and other plant material by soil microorganisms.

**Hydrogen Bond:** There is an electrostatic attraction between the oxygen atom of one water molecule and hydrogen of another water molecule which is known as hydrogen bond.

**Hydrophilic:** Compounds that readily dissolves in water are referred to as Hydrophilic (Greek, "Water-loving").

**Hydrophobic:** Compounds that do not readily dissolve in water are termed as Hydrophobic.

**Hypoxia:** Hypoxia is a condition where not enough oxygen makes it to the cells and tissues in the body.

**Indicator:** The end point or equivalent point of titration is recognized with the help of a chemical reagent called "Indicator".

**Lethal effect:** Results in death of an organism

**Magma**: It is the molten or semi-molten natural material from which all igneous rocks are formed.

**Mineralisation:** It is defined as conversion organic compounds into inorganic compounds through various decomposition procedures.

**Molality:** It is the number of moles of solute per kilogram of solvent. Solutions expressed in molal concentration are denoted with 'm'. 1 m solution contains 1 mole of solute per kilogram of solvent.

**Molarity:** It is a unit of concentration measuring the numbers of moles of a solute per litre of solution. Molarity is denoted with 'M'. A 1 M solution contains 1 mole of solute per litre of solution.

**Mulch**: A mulch is a layer of material applied to the surface of soil.

**Nephrocalcinosis:** Formation of calcareous deposits in the kidney.

**Normality:** It is a measure of concentration equal to the gram equivalent weight per litre of solution. Gram equivalent weight is the measure of the reactive capacity of a molecule. The solute's role in the reaction determines the solution's normality. Normality is also known as the equivalent concentration of a solution.

**Parts per million (ppm):** A concentration of 1 ppm means 1 mg of solute in 1,000,000 mg of solution, or 1 mg of solute in 1000 g of solution, or 1 mg of solute in 1 kg of solution. This also means 1 mg of solute in 1000 mL or 1 L of solution (remember that 1 mL of water weighs 1 g or 1 L of water weighs 1 kg).

**Parts per thousand (ppt):** A concentration of 1 ppt means 1 mg of solute in 1000 mg of solution, or, 1 g of solute in 1000 g (1000 mL or 1 L) of solution.

**Peat soils:** Peat soils are formed from partially decomposed plant material under anaerobic water saturated conditions.

**Pedogenesis**: The process of formation of soil through weathering and natural erosion of the parent material is called soil formation or pedogenesis.

**Pedology:** The study of soil formation is called pedology.

**pH:** pH is defined as negative logarithm of $H^+$ ion concentration.

**Physical properties**: Physical properties of water are related to the appearance of water, namely, the colour, temperature, turbidity, taste, and odour.

**Poikilotherm:** These are ectotherms which have no control over their body temperature; their core body temperature conforms to ambient temperature.

**Saline soil:** Saline soil is defined as one that inhibits the growth of most crops because of the higher concentration of neutral soluble salts in it.

**Salts:** Salts are the compounds obtained by neutralization of acid and base.

**Sample:** A sample is defined as a portion of material selected from a larger quantity of material.

**Soil amendments:** These are organic or inorganic matter added to the soil to improve texture, water retention, drainage or aeration.

**Soil Chemistry:** Chemical constituents, chemical properties and the chemical reactions of soil.

**Soil deodorizer:** These are agents used for suppressing the explosive generation of spoilage and decaying odors from soil.

**Soil fertility:** Nutrient supplying properties of soil.

**Standard solution**: Standard solution is the one whose concentration or strength is accurately known. It is also sometimes referred to as "titrant".

**Stenohaline:** Species that can thrive only on a narrow range of salinity.

**Stress:** The general physiological response of fish to threatening situations, as with all vertebrates, is referred to as stress.

**Sub-lethal effect:** Effect not enough to cause death but other health problems in an organism.

**Tilling:** Tilling is the act of preparing the soil for planting.

**Titration:** The Process of adding a standard solution from a volumetric burette to a solution of the unknown concentration in the conical flask until the reaction is just complete is known as "Titration".

**Total Solids:** Total solids in the water sample represent the dissolved organic matter, particulate organic matter, and dissolved inorganic substances.

**Turbidity:** Turbidity refers to an optical property of water that causes light to be scattered or absorbed rather than transmitted through the water in a straight line.

**Volumetry:** Volumetry is measurement of volume of a solution of known concentration which is used to determine the concentration of the analyte.

**Warmwater fish species**: Tropical fish species whose congenial temperature range is 24-30°C are referred to as warm water fish species.

**Weathering:** Weathering is a term which describes the general process by which rocks are broken down at the Earth's surface into sediments, clays, soils and substances that are dissolved in water.

**Wet lands:** Surfaces which remain under water most of the time.

# References

APHA, 2004. Standard methods for the examination of water and wastewater, American Public health Association Washington *D.C.*1134p.

Barman, U. K., Jana, S.N., Garg, S. K., Bhatnagar, A. and Arasu, A.R.T., (2005), Effect of inland water salinity on growth feed conversion efficiency and intestinal enzyme activity in growing grey mullet, *Mugil cephalus* (Lin.): Field and laboratory studies, *Aquaculture international*, 13(3), pp 241-256.

Bhatnagar A, Jana SN, Garg SK, Patra BC, Singh G, Barman UK. Water quality management in aquaculture, In: Course Manual of summer school on development of sustainable aquaculture technology in fresh and saline waters, CCS Haryana Agricultural, Hisar (India), 2004, 203- 210.

Boyd, C.E. & Tucker, C.S., 2015. Pond aquaculture water quality management (2$^{nd}$ edition), Springer Science Business media, 700p.

Glasser, L & Doerfler, R (2018). A brief introduction to nomography: graphical representation of mathematical relationships. *International Journal of Mathematical Education In Science And Technology* 50(8),1273-1284 https://doi.org/10.1080/0020739X.2018.1527406

Helmenstine, Anne Marie, Ph.D. Acid Definition and Examples. Thought Co, Aug. 29, 2020, thoughtco.com/definition-of-acid-and-examples-604358.

ICAR, New Delhi, 2011.Handbook of Fisheries & Aquaculture, DKMA, ICAR, New Delhi, 1116.

Kolay, A.K., 2007. Soil genesis, Classification Survey and Evaluation (Vol.2), Atlantic Publishers & Distributors (P) Ltd., 744p.

Manahan, S.E., 2000. Environmental Science, Technology, and Chemistry; Environmental Chemistry, first ed. CRC Press.

Ramdhas, V. & Santhanam, R. (1996) A manual of Methods of Seawater & Sediment Analysis, 1st Edition, TNVASU, Chennai.

Latha, M.R. & Janaki, P., *Problem soil & their management*, TNAU, Coimbatore, 102p.

Jain R., Saxena J. (2019) Impact Assessment of Microbial Formulations in Agricultural Soil. In: Singh D., Gupta V., Prabha R. (eds) Microbial Interventions in Agriculture and Environment. Springer, Singapore. https://doi.org/10.1007/978-981-13-8383-0_16

Jain, R., & Saxena, J. (2019). IImpact Assessment of Microbial Formulations in Agricultural Soil. Microbial Interventions in Agriculture and Environment. Vol. 2: Rhizophere, Microbiome and Agro-ecology, 471-495.